BIOMECHANICS - THEORY AND APPLICATIONS SERIES

ARTS BIOMECHANICS – AN INFANT SCIENCE: ITS CHALLENGES AND FUTURE

BIOMECHANICS: THEORY AND APPLICATIONS SERIES

Biomechanics: Principles, Trends and Applications
Jerrod H. Levy (Editor)
2010. ISBN: 978-1-60741-394-3

Arts Biomechanics - An Infant Science: Its Challenges and Future
Gongbing Shan and Peter Visentin
2010.ISBN:978-1-60876-809-7

BIOMECHANICS - THEORY AND APPLICATIONS SERIES

ARTS BIOMECHANICS – AN INFANT SCIENCE: ITS CHALLENGES AND FUTURE

GONGBING SHAN

AND

PETER VISENTIN

Nova Science Publishers, Inc.

New York

For permission to use material from this book please contact us:
Telephone 631-231-7269; Fax 631-231-8175
Web Site: http://www.novapublishers.com

LIBRARY OF CONGRESS CATALOGING-IN-PUBLICATION DATA

Shan, Gongbing.
Arts biomechanics : an infant science : its challenges and future / Gongbing Shan, Peter Visentin.
p. cm.
Includes index.
ISBN 978-1-60876-809-7 (softcover)
1. Human mechanics. 2. Arts medicine. 3. Arts--Physiological aspects. I. Visentin, Peter. II. Title.
QP303.S488 2009
612.7'6--dc22

2009045658

Published by Nova Science Publishers, Inc. † New York

CONTENTS

Chapter 1	**Arts Biomechanics – An Infant Science: its Challenges and Future**
Chapter 2	**Performance Skill Analysis and Acquisition**
Chapter 3	**Injury Risk Quantification and Prevention Related to Artistic Performance**
Chapter 4	**Applications of Movement Science Technologies in the Creation of Art**

Acknowledgements
References
Index

Chapter 1

ARTS BIOMECHANICS – AN INFANT SCIENCE: ITS CHALLENGES AND FUTURE

1.1. OVERVIEW

While biomechanics has achieved successes in many fields involving locomotion, motor learning, skill acquisition, technique optimization, injury prevention, physical therapy and rehabilitation, one area has heretofore been scarcely represented in the literature – Arts Biomechanics. Biomechanics clearly has significant potential for application in the performance arts, such as music and dance, since skills needed for these activities are visibly related to the human musculoskeletal and nervous systems. In such areas, Arts Biomechanics should begin by focusing on skill analyses and acquisition necessary for the performance of the artistic act. Subsequently it should engage in a deeper discourse that explores the relationship between these and the desired aesthetic outcome. Less apparently, biomechanics may also enhance the analysis and comprehension of other arts, such as painting, where gesture is often embedded in the artwork by means of symbolism, tradition, the process of art creation, or as an inherent product of the existential nature of humanity.

There are many challenges facing the integration of the Sciences with the Arts. On a fundamental level, the principles and goals of one often seem at odds with the other. In reality, neither science nor art is antithetical to the other.

> "The most beautiful experience we can have is the mysterious. It is the fundamental emotion that stands at the cradle of true art and true science. Whoever does not know it and can no longer wonder, no longer marvel, is as good as dead, and his eyes are dimmed. It was the experience of mystery -- even

> if mixed with fear -- that engendered religion. A knowledge of the existence of something we cannot penetrate, our perceptions of the profoundest reason and the most radiant beauty, which only in their most primitive forms are accessible to our minds...I am satisfied with the mystery of life's eternity and with a knowledge, a sense, of the marvelous structure of existence -- as well as the humble attempt to understand even a tiny portion of the Reason that manifests itself in nature"(Einstein, 1931).

However, the self-perceptions of artists and scientists may be problematic. Scientists take reductionist and reasoned approaches to the world: there is a position/argument; it is structured in logical steps; a topic sentence names and proves each idea in the discourse; and, the discourse is aimed at a specific result. For the scientist, these give feelings of comfort, power, and/or control over the phenomenon. Conversely, the artist is less interested in the factual and desires to convey the emotional and sensual; reductionist, topical arguments are considered antithetical to the creation of good art; too precisely formulated a conceptual structure is perceived as negating mystery and limiting artistic possibilities; and, the result is a product of the moment – an ever shifting target in time-based performance art. Repeatability is only a desirable quality in that it can be a measure of skill level. Hence it shows little in terms of creative ability. The artist wants power and control over others' perceptions of the phenomenon; each individual audient's experience of the art responding to and thus validating the artwork and, by extension, the artist. Perhaps it is the way in which scientists and artists perceive their own roles that a binary view pervades their respective disciplines – something is or isn't. Given this seeming dichotomy, the question arises "Why bother with Arts Biomechanics?"

The issue that provides the main inertia to expand this nascent field of biomechanics comes from artists. It is a medical one. Epidemic rates of debilitating injury (48-76%) occur in performing arts such as music and dance (Brown, 1997; Fry, 1986, 1987, 1988; Fry, Ross, and Rutherford, 1988; Hagglund, 1996; Hartsell and Tata, 1991; Lockwood, 1988; Middlestadt and Fischbein, 1989; Zaza, 1992, 1998). Unfortunately, there is currently little quantitative research examining the aetiology of performance injuries. Existing strategies to address injuries are largely qualitative and "experience" based. And they are normally employed only after injury has occurred. Even as artists are becoming increasingly aware regarding their physical needs, career longevity and injury downtime are becoming pervading arts industrial issues. Only recently have researchers begun to explore scientific approaches, such as human performance engineering, to trace causal factors related to human bone, muscle, and nervous systems injuries in the performing arts (Chesky, Kondraske, Henoch, Hipple, and

Rubin, 2002; W. J. Dawson, 2003, 2007; Solomon and Solomon, 2004). Finally, artists themselves are slowly turning toward science to provide preventative answers and not merely remedial ones.

From a phenomenological point of view, the performing arts share many characteristics, including health risks, in common with other skill-oriented activities (Chesky et al., 2002; Lehmann and Davidson, 2002; Wilson, 1986). Commonalities between athletic and artistic performance seem obvious, particularly in the area of motor skill analysis, acquisition during skill learning and performance. Like athletes, elite musicians practice, with many hours of repetition, to perfect complex motor control sequences. However, music students are seldom introduced to basic principles of movement science and physiology that underpin that activity. This is largely because the focus of music teaching is artistic and outcome driven rather than process oriented – this is a difference between practicing and training. While some teachers are knowledgeable regarding efficient use of the body for the benefit of good performance, many are not. In many parts of the world, classical musicians today are taught in a manner virtually indistinguishable from that used 50 years ago. Nowhere in the world could the same be said for elite sport.

As such, it seems logical to build on the successes of Sports Biomechanics in the service of the Arts. Motor learning, skill acquisition and learning while minimizing injuries constitute the main research focuses of Sports Biomechanics (Ballreich and Baumann, 1996). However, there is a fundamental ethical difference between these two fields. Whereas Sports Biomechanics typically directs is energies to achieving specific, goal-driven, quantifiable results that are valued for their repeatability (e.g. faster, higher, stronger, etc.), Arts Biomechanics must be satisfied with guiding the process without appearing to identify an absolute goal – for reasons of artistic and creative freedom. The emphasis of each is on training, but Arts Biomechanics needs to contribute to a demystification of the learning and skill acquisition processes, so that artists can realize their full potential *and* survive their chosen vocation.

> "My early love affair with dance gradually had been replaced by a struggle to become what I could not be. I had selected teachers who felt it was their responsibility to tell me over and over what was wrong, rather than helping me generate the knowledge that would give me tools to make things right" (Evans, 2003).

Clearly, the need to consider artistic values as well as scientific ones creates particular challenges for the field of Arts Biomechanics. Facing these challenges is, in our opinion, best served by a multidisciplinary approach – one where research does not simply adopt a scientific or artistic practice, but engages in discussion that focusses on liminal spaces of understanding among disciplines, ultimately transcending the viewpoint of each. Meaningful and relevant research results will be those where science informs artistry rather than attempting to modify it. Currently, music and dance are the two dominant areas of biomechanics research in the arts. For this reason, discourse below will primarily focus on these disciplines. The main effort of this book is to summarize the state of Arts Biomechanics in the following areas: 1) skill analysis, acquisition and pedagogy, 2) injury risk identification, quantification, prevention and compensation strategies, and 3) Innovative uses of the tools of movement science in the analysis and creation of art. Further, this book will provide discussion that identifies some of the challenges facing Art Biomechanics, elaborate on its potential and identify some future directions.

Chapter 2

PERFORMANCE SKILL ANALYSIS AND ACQUISITION

2.1. HISTORICAL OVERVIEW

Documentation pertaining to instrumental performance and dance has a long history. In terms of modern classical performance traditions, documentation that incorporates information on mechanics and motor skills begins as early as the 16th century for instrumental performance and the 17th century for dance (Hilton, 1997; Kolneder, 1993). Its very existence can be considered evidence of a general desire to improve overall quality of music/dance performance, both stylistically and technically. Most of the documents are pedagogical in nature and they are clearly not targeted at the mature or virtuoso performer. For example, in his seminal treatise on learning to play the keyboard, Carl Phillip Emanuel Bach clearly states his intent.

> "...keyboard instruction could be improved in certain respects to the end that the truly good which is lacking in so much music, but particularly keyboard music, might thereby become more widespread. The most accomplished performers, those whose playing might prove instructive, are not to be found in such numbers as might perhaps be imagined" (Bach, 1759).

In terms of classical dance, the beginnings of ballet and classical notation are associated with the Court of Louis XIV of France (1639-1714) (Hilton, 1997). Iconography, dance manuals and descriptions of the desired aesthetic provide some of the first rudimentary (bio)mechanics information for analysis and acquisition of dance skills (Figure 1 and 2).

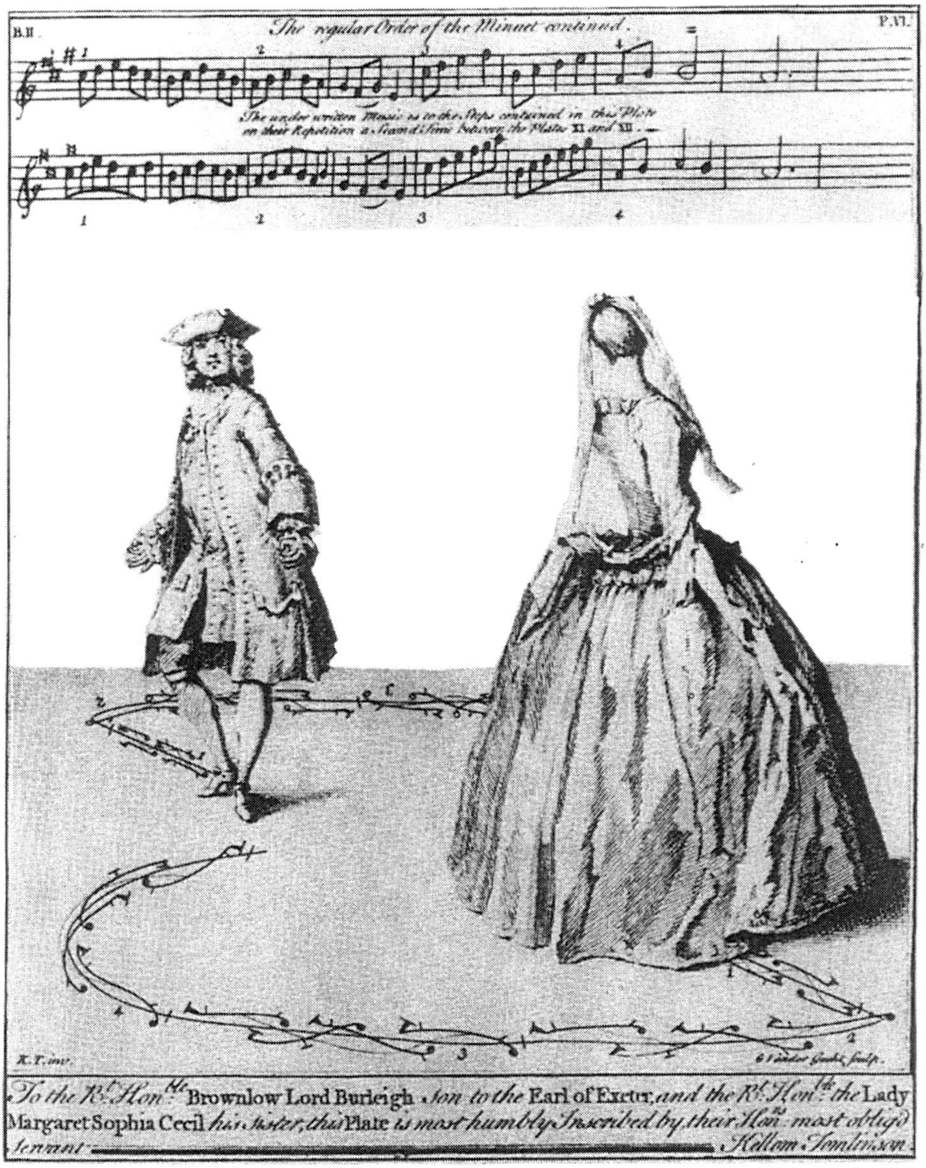

Figure 1. Choreography of a Minuet providing rudimentary instruction on placing the feet. The original source is from an English dance manual by Kellam Tomlinson, and documents the influence of French culture in the 18th century.

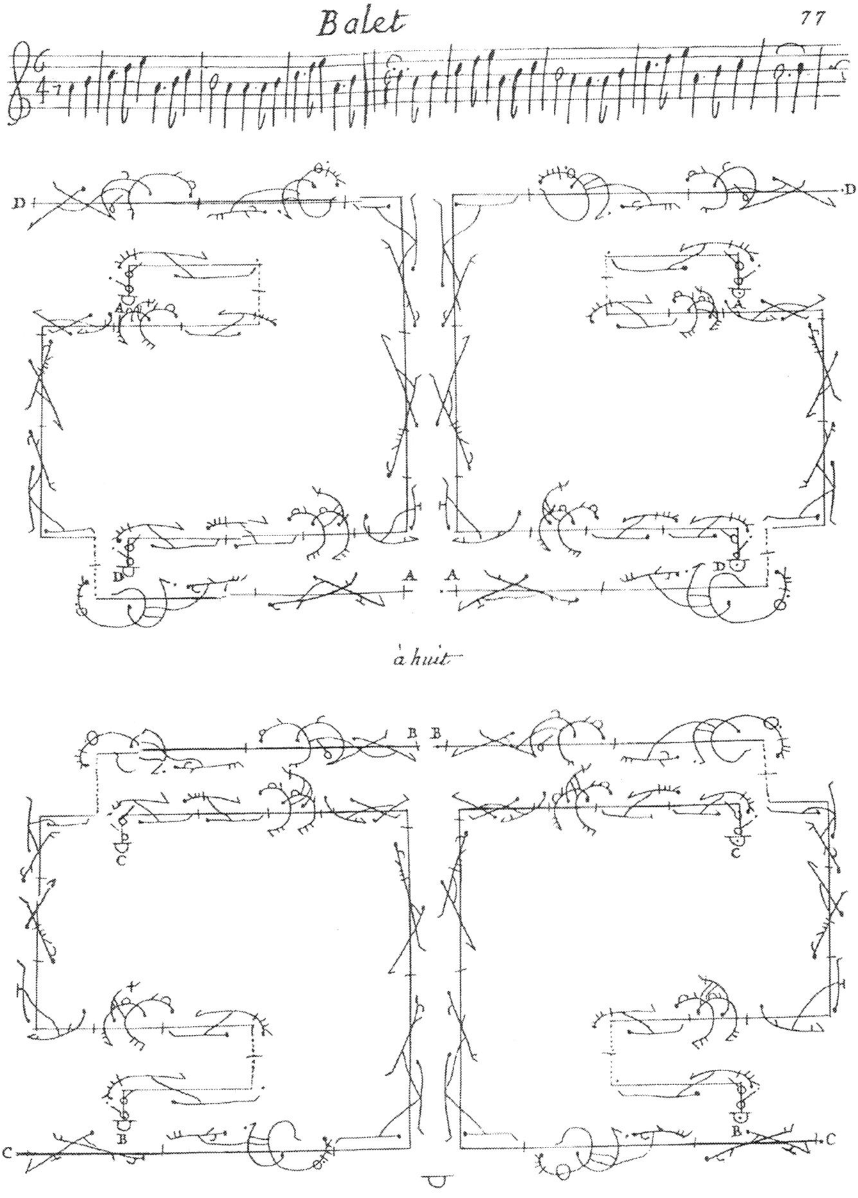

Figure 2. A choreography of steps from "Balet de Neuf Danseurs" by Feuillet. Paris, 1700.

Collectively, such sources contain fundamental instruction on music/dance and their contemporary aesthetic goals. They include postural descriptions, instruction pertaining to skill acquisition such as: fingerings in basic positions, embouchure (for wind instruments) and bowing techniques (for strings), and they describe the phenomenology of physically interacting with a musical instrument (how to hold it, the consequences of certain muscular tensions, tablatures revealing mechanical insights, etc.). Choreographed movement in dance also creates a gestural map for musicians as well as dancers, since the music must maintain an aesthetic sensitive to bodiliness required in dancing (Bach, 1759; Hilton, 1997; Mozart, 1756; Quantz, 1752; Tromlitz, 1791). Clearly, these sources point to an awareness of the mechanics of the human-tool interface, whether that interaction is with an instrument or a physical space.

Further, almost on a century-by-century basis, one can observe a progression to more and more systematic approaches toward artistic performance. Documentation pertaining to violin performance can be considered exemplary in this regard. Earliest documentation, from the 16th century, was descriptive, providing relatively little instruction on technique. The 17th Century included some of the earliest tutors. They contained fundamental postural descriptions and instruction on music basics: string tunings, fingerings in basic positions, elementary bowing techniques. Included are some instructions on bow direction, accents, and attempts to describe articulations and expression. Tablatures (musical fingering charts) revealed insights into pedagogy. Information on style and technique may also be gathered from music that incorporated performance indications by the composer. The 18th Century rise of a middle class and a corresponding increased interest in education lead to maturation of instruction methods and pedagogy. More methods and music were published. Contents became more detailed with broader variety of musical examples and more instructions how to coordinate techniques in order to play expressively. Some compositions dealt with specific technical difficulties or material for specific skill acquisition.

The 19th Century showed evidence of a paradigm shift (Baillot, 1835). There was a move from general education and a "collective consciousness" to a focus on individual training and virtuosity. The period marked the beginnings of music pedagogy as a science. Documentation used a new tone that emphasized the development of technique as a means to achieve the highest degrees of artistry. The focus was on technical training (Figure 3). During the period, a gradual increase in notion of finding the "best" way to practice can be observed. Some writers advocated a "mechanical approach" to the instrument with attempts to apply scientific methods to describe these mechanics. Others began to approach the subject of performance psychology. Regarding anatomy and physiology, methods to this point may be summarized as rudimentary. They were experience/practitioner-based with superficial descriptions of postures or (bio)mechanics (Baillot, 1835). Attempts to include an understanding of anatomy and physiology were a development of late 19th and early 20th centuries.

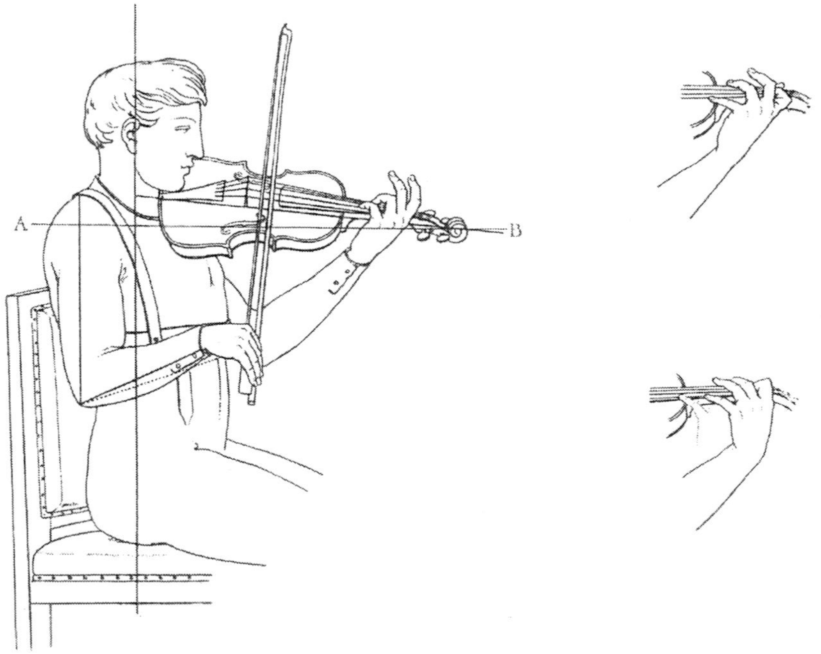

Figure 3. Position of the right arm and wrist held closer to the body when the violinist plays sitting down. The neck of the violin is then held slightly lower. Position of the hand, wrist and fourth finger in extensions (top-right). Forced and improper position of the hand, wrist and fourth finger in extensions (bottom-right).

Over the course of the 20th century, there was a continued development of systematic teaching methods. Also, there was an increasing trend toward learning from the standpoint of different disciplines: physiology, psychology, (bio)mechanics. Some training exercises were developed (at times by individuals who may have had little to no "practical" experience playing music – e.g. medical doctors, physiology professors) that pushed mechanics to conceptual and physiological limits. As can be seen in Figures 4-6, the body was typically treated as a machine which could be trained, or programmed, through repeated and structured movement variations (Hodgson, 1958).

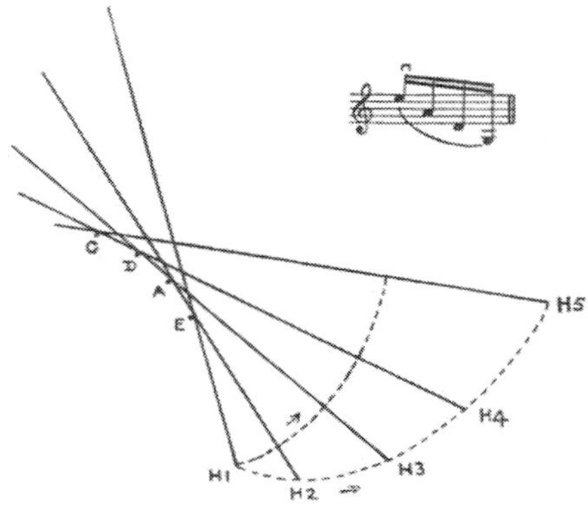

Hodgson, 1958.

Figure 4. Illustration considering the curve which results from the identified ordering of notes, while using a down bow. The two dotted lines give variations in the track followed by the hand.

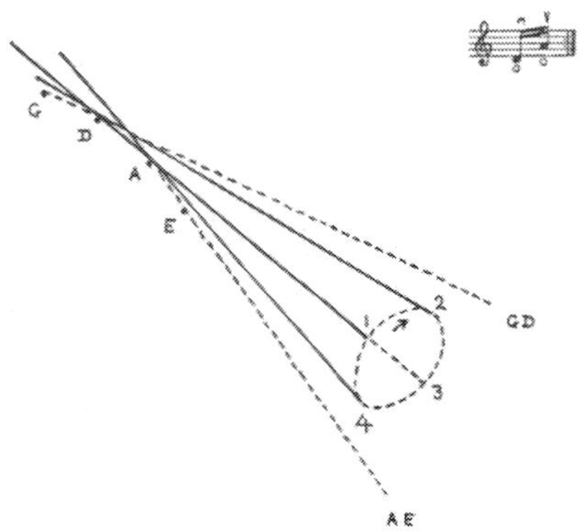

Hodgson, 1958

Figure 5. The figure shows a possible elliptical path (marked 1, 2, 3, 4), traversed by the heel of the bow during the playing the identified two notes.

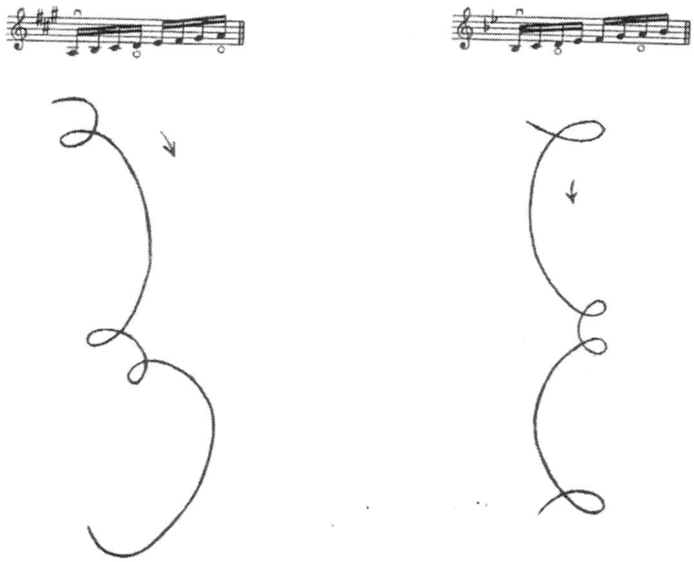

Hodgson, 1958

Figure 6. Small loops occurring in the bow-hand that result from string crossings for the identified scales. String crossings occur on an "up-bow" for the figure on the left and on a "down-bow" for the one on the right.

Most 20th-century writings that attempted to document teaching methods of "successful" practitioner/pedagogues failed to deal with fundamental psyco-physiological learning, biomechanics, or neural control except in superficial terms (G. B. Shan, Visentin, Wooldridge, Wang, and Connolly, 2007). For example, after mentioning postural or internal motor control factors in general terms, most typically continue with experience-based description of techniques and their application in selected passages of repertoire. Thus, they are written in such a way that only those who have experience with the phenomenon and the aural tradition being described can sympathize with the descriptions.

Clearly, in the 21st century, there can be a role for the field of biomechanics in terms of the pedagogical process. Demystification of the mechanics of fundamental skills, and direction regarding the best way to learn them, is in the best interests of these disciplines – especially if it can lower injury rates and/or improve the physical health and well-being of performing artists. Currently, Arts Biomechanics in its infancy as a field of study.

2.2. Challenges Faced by Arts Biomechanics

Notwithstanding efforts to address important artistic concepts through written documentation, the performance arts has and continues to perpetuate its traditions mainly through an apprenticeship learning model. Advanced teaching/learning typically occurs in a one-on-one environment and arts education relies heavily on aural traditions, with pedagogical frameworks by and large limited by conventions and aesthetic values passed from teacher to pupil. Participants in such a model typically take great pride in the provenance of their traditions – often validating their approach by tracing their artistic lineage from teacher to teacher to one or another of the legendary "greats" of the past. Such a mind-set frequently inspires a mystical and, in some cases, a near-religious reverence of the teacher, artistic traditions, concepts, and acquired skills. Nearly everyone who is involved in the performing arts can, through personal experience, cite anecdotal evidence confirming this assertion.

Research in Arts Biomechanics can play an important role in challenging the inertia of tradition that exists in the performing arts. By supplying quantitative rather than empirical evidence for improving pedagogy, this sub-discipline of applied biomechanics can improve understanding of physiological phenomena, motor control, the motor learning process and internal loading implications for the body during practice and performance. With care, it can do so without being perceived as a threat to the artistic end-product.

In such a system, and since the performative arts are heavily influenced by the need for an audience, *performance* and *pedagogy* are driven by external validation. Frequently, a teacher's quality is judged by the successes of his/her "prize" students. As artistic success or failure is often determined by micro-increments of "talent", this can lead to an emphasis on immediacy of skill acquisition, sometimes at the expense of sound learning practices (Visentin, Shan, and Wasiak, 2008). According to Ruggiero Ricci, one of the 20th century's premier violinists:

> "Teachers are ruining a lot of kids. Parents love to say 'my kid plays the Mendelssohn Concerto' so they take their child away from a teacher who knows better and gives him to one who will push him. ... they are the real criminals in this business" (Schwartz, 1983).

Notwithstanding this somewhat arcane approach, numerous examples of success within the model reinforce the perceived legitimacy of its practices as the best one available to cultivate "talent".

However, there are significant problems with such simplistic attitudes and rationales. First, in order to be scientifically defensible, the existence of talent must be proven to be either an innate characteristic of individuals, somehow defined by their genetic makeup, OR a byproduct of learned behaviors and environmental circumstance. At any given point of time, it is relatively easy to observe levels of skill development and ease of skill acquisition (sometimes equated to talent) for individuals within a group of learners. In one scientific study, subjects were sorted into two groups based on expert assessments of their "talent" (Levitin, 2006). After following the groups for several years, results proved discrimination on the basis of talent to be a false determinant of success over the long term. Learning/practice time was the ultimate cause of achievement, and not innate ability. Hence, a focus on the learning process needs to dominate arts pedagogy.

Further, success as defined by the performance is a measure which oft-times fails to concern itself with the long-term health and development of the artist. Since the medical literature shows overwhelming evidence of vocational disaster over the "long-haul" due to technique-related injuries, there has been an increased interest in improving pedagogical practice in the performing arts. Past studies have shown that 48% - 76% of professional musicians need to stop performing for extended periods due to occupational injury (Brown, 1997; Fry, 1986, 1987, 1988; Fry et al., 1988; Hagglund, 1996; Hartsell and Tata, 1991; Lockwood, 1988; Middlestadt and Fischbein, 1989; Zaza, 1992, 1998). As a result, old platitudes such as "the way to Carnegie Hall is practice, practice, practice" have evolved into "the way to Carnegie Hall is practice, *rest*, practice" (Kaplinski, 2003). Further, traditional practices, such expecting students to engage in long hours of repetitive work often with little consideration for its efficiency or biological implications, are now becoming recognized as part of the problem, even if there is a reluctance to changing traditional mind-sets. In summary, the performing arts appear to be ready for more collaborations and science has a great deal to offer.

Arts Biomechanics can play an important role in challenging the inertia of tradition that exists in the performing arts. By supplying quantitative rather than empirical evidence for improving pedagogy, this sub-discipline of applied biomechanics can improve understanding of physiological phenomena, motor control, the motor learning process and internal loading implications for the body during practice and performance. And, with care, it can do so without being perceived as a threat to the artistic end-product. *Ultimately, this is the most significant practical challenge for Arts Biomechanics.*

Unfortunately, there is currently little quantitative research in Arts Biomechanics. One reason for this may be that both the performing arts and biomechanics are highly specialized studies each with their own cultures. These cultures may have language barriers (the same words can have different disciplinary meanings, and hence loading), goal differences (personal and subjective experience is desired for one and quantitative observation of a process that leads to specific goals for the other), and/or conflicts between their needs (the arts require flexible outcomes that accommodate changing tastes/ideas and the sciences desire outcomes that are essentially quantifyable/immutable). Further, from the standpoint of arts pedagogy, because of anthropometrical variation between teacher and student, it can be the tendency for an arts teacher to focus on stylistic outcomes, which in turn obliges the student to find their own motor control means to that particular end.

Arts Biomechanics can inform the teacher/pupil interface, supplying common ground so that a teacher can adjust his/her understanding to a student's individual anthropometry, and so that the student can better comprehend the teacher's instruction. Biomechanics has great potential to provide linkages between science and the arts. Clearly, collaborative work is needed in order to bridge disciplinary divergences. In short, Arts Biomechanics needs to be an integrative force, a nexus for communication that is not merely coordinative of differing views but accounts for both perceptual experience and scientific quantification at all stages of the work. *This is the most significant conceptual challenge in the application of biomechanics to the arts.*

Finally, the overview of this chapter began by identifying some philosophical challenges for Arts Biomechanics. Individuals within the arts and the sciences have a vested interest in maintaining the fundamental precepts and traditions which define the relationship between discipline and practitioner. For the artist, quantification and transparency represent threats to individuality and artistry. For the scientist, subjectivity is something that merely has not yet been objectified. *Changing attitudes to foster recognition of the potentially symbiotic relationship and finding convergences that can link these ethics are the most significant philosophical challenges for Arts Biomechanics.*

In any motor skill endeavor, skill acquisition can be categorized into three primary stages: cognitive, associative and automotive (Magill, 2001). Clearly, given the cultural characteristic of the performing arts and applied biomechanics disciplines, the greatest potential for collaboration between the two seems to lie in addressing the cognitive and associative stages of the learning process. Such an approach does not challenge artistic values, and it builds on the strengths of biomechanics as a means to understand and further motor skill development. Additionally, it has potential to effectively address injury prevention through quantitative research examining the aetiology of performance injuries. Since most existing strategies to address injuries are largely "experience" based and they are normally employed only after injury has occurred, it appears logical to begin with an examination of pedagogy's contribution to this problem. Arts Biomechanics has an important role in this regard (Visentin et al., 2008). To this end, Arts Biomechanics as a field of study has begun with the following two questions in mind: "can we identify, describe, and *quantify* qualities that lead to artistic success while minimizing the potential of performance injury?" and "how do we communicate these qualities broadly to the appropriate audience?"

2.3. Arts Pedagogy; What Can We Learn From Biomechanics Applications?

The use of biomechanics-based motor learning in sports pedagogy has proven highly successful in improving human performance (Ballreich and Kuhlow, 1980; Magill, 2001). Its focus has usually also been on enhancing physiological limits of athletes. Superficially, sport and the performing arts may appear to have more or less in common, depending on the art. For example, the athletic aspects of dance are obvious whereas those of music performance often are not. All rely upon the training of human motor skills and the development of gross and fine motor control. Fundamentally, both sports and arts performance involve motor learning and the sublimation of neural control to effect a desired physical outcome. But, in music and dance the focus is not overwhelmingly on pushing physiological limits as it is in sports. Rather it is about controlling the process during the performative act. If we continue with a sports analogy, in this regard arts performance can be likened more to acrobatic kinds of sports than to those which rely mainly upon speed and power. In much of arts performance, fine motor control seems to be a necessarily dominant concern. In music, although success is judged by an "aural" outcome, it is a product of physical gestures. In dance, the result is a visual one that is obviously guided by gesture. Since sports biomechanics is at an advanced stage of application, it seems sensible to adopt and modify its already successful methodologies to the context of the performing arts. The perspective it can provide may contribute to a general demystification of the teaching/learning process in the arts with a corresponding improvement in overall pedagogical success as measured by both learning outcomes and injury prevention. When adapting a sport-based motor-learning model to the arts, one must consider its multidisciplinary character.

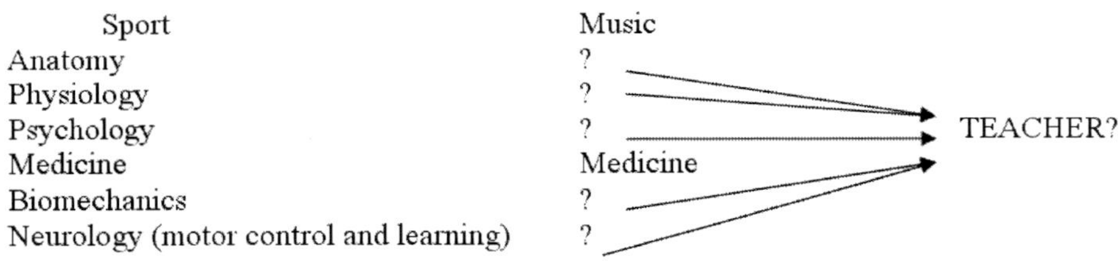

Figure 7. A comparison of the resources used in the pedagogical process in sport and music

In sports, intervention of related disciplines to influence pedagogy and learning is a mature science; a team of specialists guides training and learning. Interdisciplinary research is commonly understood to hold significant potential for improving performance. In performing arts, educationional resources are considerably fewer. For example, in music responsibility rests primarily, if not solely, with the teacher who may only have minimal training in some of the multiple required areas of competency (Figure 7). Even in the area of arts medicine, prevention is an underutilized strategy for artists. Mostly, medical attention is only sought for injury remediation.

The consequences of poor health and wellness practices for performers can be devastating. They include pain, injury, long-term disability, and even career-ending pathologies. With regard to the intervention of related disciplines, arts pedagogy is in a state of development similar to that of sports pedagogy about fifty years ago. Learned behaviours in the arts are dominated by the vision of the end result. In this regard it is fair to draw from sport history (fifty years ago) as accurately representing and predicting the needs in the arts. One sports anecdote that universally receives sympathy, acceptance and affirmation when presented to the performing arts community involves Tom Simpson, British cyclist and world professional road racing champion in 1965. His mind-set can be summarized by his two most famous quotes. An advocate of performance enhancing drugs, he is noted as saying “if ten will kill me, give me nine”. And his final words as he lay dying from cardiac arrest at the top a mountain during a race were “put me back on my bike” (*Fotheringham, 2002)*. Arts practitioners empathize with this “pushing the limits” attitude.

Accordingly, the first aims of those who investigate Arts Biomechanics have been to begin exploring the utility of methodologies and technologies already successfully employed in mature fields of kinesiology and sports medicine to examine arts performance. The bases for most of these explorations have been grounded in two fundamental practices of kinesiology: motor control and motor learning. This approach provides a methodological means to analyze movement, cognition and the learning process, and it supplies an improved basis for discourse through an established commonality of idiom and terminology. Simply, it allows us all to speak the same language.

A Motor-Learning Model

Two foundational theories provide important perspective on skill acquisition in the performing arts. The first, Ballreich's model for motor learning identifies three requisite elements for success: (a) accurate characterization of the skill to be acquired, (b) skill transfer between dissimilar bio-systems, and (c) skill acquisition without injury (Ballreich & Baumann, 1996).

1. *Characterization of the skill.* Characterizing the skill to be acquired involves scientific analysis to identify and describe motor control patterns (i.e. the interaction and coordination of the neural and musculoskeletal systems). Methodologies from kinesiology such as motion analysis are ideally suited to this purpose. Motor control patterns common among expert performers can be generalized, establishing "markers" which can be used as references in the pedagogical process. Using this approach, characteristics of high-level performers can be contrasted with those observed in learners. Such a quantification of fundamental skills begins to address the most practical challenge in the arts – a clear understanding of motor control patterns needed to communicate the desired outcome.
2. *Skill transfer between dissimilar bio-systems.* For a successful motor learning model, fundamental skills must be understood in such a manner that the teacher can transfer the principles to the student regardless of physical differences. The most significant pedagogical weakness in "experience-based" teaching approaches is the fact that not everyone is the same anthropometric shape. Simply, the teacher's experience does not *really* apply. Clear characterization of skills and marker-based learning can facilitate skill-transfer between teacher and student by providing a general model of the specific kinematic gestures vital to skill acquisition.

However, for successful skill transfer, "universal" characterizations need to adjust for the "particular", compensating for anthropometric differences among learners. For this to occur, a sufficient body of Arts Biomechanics research is required to inform teachers' practical experience.

3. *Learning without injury.* By directing attention to specific motions or neural control patterns that novices must assimilate into their technique, skill transfer can occur efficiently and effectively; reducing the assimilation time needed to acquire each skill or technique, and thus reduce the risk of Repetitive Stress Injuries.

The second theory addresses skill development at the different stages of the learning process. Relatively simple, the Fitts and Posner model has been proven effective in the discipline of kinesiology (Magill, 2001). It identifies three general stages: cognitive, associative and autonomous.

1. *Cognitive.* The beginner focuses on cognitively oriented problems. Performance is highly variable, showing a lack of consistency from one attempt to the next. Often, learners are confused about what they need to do in order to improve.
2. *Associative.* Cognitive activity changes because the learner has begun to associate multiple references in order to guide the outcome. Through training and repetition learners begin to make fewer gross errors as they refine and combine the multiple skills acquired in the cognitive stage of the process.
3. *Autonomous.* Here the skill becomes almost automatic or habitual. Learners can perform with repeated degrees of accuracy, without conscious thought.

Both of these models address different aspects of motor learning; one deals with biomechanical factors influencing the process of motor learning and the other concentrates on the process of skill development (Ballreich & Baumann, 1996; Magill, 2001). Collectively, these two models have important implications for the development of pedagogical practices in the arts. Figure 8 is a combinatorial model, representing Ballreich's model for motor learning on the vertical axis and Fitts & Posner's skill development model on the horizontal one (Visentin et al., 2008).

In our opinion, rationalizing this model for arts pedagogy requires careful consideration and a nuanced understanding of disciplinary discourse within the arts. One way in which application of the model may differ from that in sports lies in the emphasis that must be placed at each stage of the learning process. Given arts pedagogy's tendency toward teleology, new ideas need significant evidence such as improved performance outcomes or faster skill acquisition rates before they will be considered. This requires: 1) clarity of definition; 2) convincing methodology; 3) consistent means of situational comparison; and, 4) solutions that do not compromise artists' perceptions of the end product. Such a rationalization of scientific methodology holds significant potential for improving arts pedagogy. But it does require finding and exploring the liminal spaces between the sciences and the arts. Hence such work can often appear to be neither science nor art.

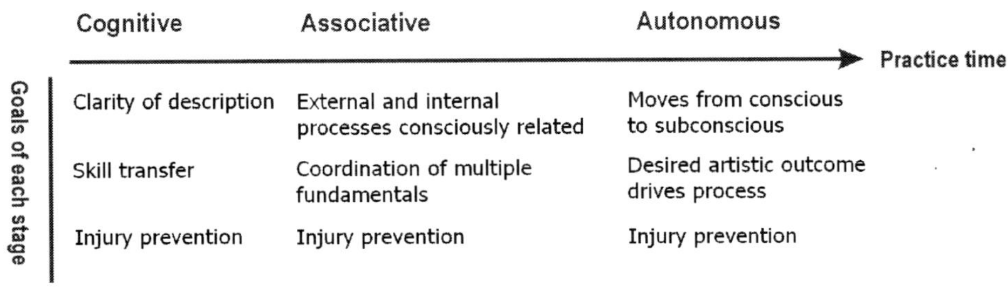

Figure 8. The stages of motor learning adapted from models of Ballreich and Magill

The combinatorial model of Fig. 8 has clear ramifications for respective roles and responsibilities of both teacher and pupil. In the cognitive stage, the onus is on the teacher to clearly and accurately characterize the skills to be acquired using instructional strategies such as modeling, clear explanations, analogies, and a vocabulary that accurately describes the phenomena. In the associative stage, teacher and student work more collaboratively toward the "particular". Compensation based on individual anthropometrics needs to be melded with aesthetic goals during this stage, as external (e.g. biofeedback instructions from the teacher) and internal processes (e.g. student's modifications to motor control according to the teacher's instructions) are consciously assimilated. Multiple skills must be coordinated together at this stage. In the autonomous stage of the process "naturalization" occurs as skill control moves from the conscious to the sub-conscious. Here, the onus for successful learning is mostly the student's responsibility and the teacher's role becomes more focused on remediation. As skill control becomes automatic, performers are freed to direct their attentions more toward aesthetic, creative and artistic outcomes – a goal that requires both forward planning and flexible response to the circumstances of the moment. The quality of instruction and practice as well as the amount of practice are important factors during all these developmental stages.

Arts Biomechanics has the potential to positively affect the performance arts through influencing the pedagogical process. It builds upon already proven scientific models of motor learning, adapting these to acknowledge artistic process and values. In the first two stages of the learning process, it can directly influence the quality of teaching/learning that occurs. This can accelerate the skill acquisition process and minimize the necessity of trial and error practice and, presumably, make the process a more enjoyable experience for the learner. It is also a motivational prerequisite, since it liberates a performer to transfer his/her attention from motor control to artistic representation, indirectly improving artists' perceptions about the third stage of the process.

Further, because arts performance involves long-term training, the dangers of vocationally related injury due to poor pedagogical practice is very high. In this regard Arts Biomechanics can highlight injury prevention as a fundamental ethic at each stage of the motor development process. It can provide realistic and practical pedagogical strategies that encourage changes in attitude and behavior. The benefits of this will be improvements to artists' health, wellness and career longevity as well as enhanced perception of their vocation.

2.4. The Tools of Arts Biomechanics Research

The scientific quantification of motor skills in arts performance is about decade old (G. B. Shan and Visentin, 2003; Visentin and Shan, 2003). Prior to this, qualitative and some quasi-quantitative image and video-based research exists in several formats; drawings, photographic images, photographic time or multiple exposures, and video (Hilton, 1997; Hodgson, 1958; Moyse, 1973; Palac, 1992; Tulchinsky and Riolo, 1994a). Existing research has primarily been concerned with skill description as it relates to artistic values/interpretation and rudimentary analyses. Most of it can be found in medical, pedagogical, and in cognitive/behavioral neuroscience (e.g. Music Perception) journals. As such, biomechanics is mostly used as an indirect or supporting means to examine disciplinary questions in these areas. As a primary area of study, Arts Biomechanics is still young. Clearly, the technologies and methods of biomechanics are an attraction to researchers from many disciplines, but the questions these researchers propose would likely be different than those a biomechanist would proffer. In Arts biomechanics, the technology must be adapted to the context of the arts. But, more importantly, the "right" research questions need to be asked; a condition that requires collaborative methodology. Since performing artists are intimately involved with their art, objectivity can best be provided by an external observer. However, because the artistic product is inherently dependent on the relationship between artist and performance, analytic objectivity has functional limitations.

A review of the most recent literature reveals that research groups looking at the performing arts are becoming more multidisciplinary. The reason for this is simple. Given the complexity of the questions being asked, it is unrealistic to expect a single discipline to provide the required artistic understanding, scientific expertise, and technological know-how. Arts biomechanics as a new area of study can directly contribute in significant ways to the latter two of these needs. Because it influences the health and wellness of performers, it indirectly contributes to artistic outcomes as well. Biomechanics brings multiple tools and scientific methodologies to bear on research which can be easily adapted to arts applications. Currently, three technologies have dominated arts performance analysis: 3-D motion capture (MoCap), force measurement, and electromyography (EMG). These technologies give rise to additional tools, such as biomechanical modeling and multidimensional signal analysis of simultaneous measurements (Visentin and Shan, 2004). The levels of sophistication currently being demanded in human movement and performance research requires synchronized information from multiple technologies in order to address more and more complex questions.

Motion Capture

3-D motion capture can be used to track movement of any object. It does so by using multiple cameras to view and triangulate the positions of markers that are placed on the item (Figure 9). Some motion capture systems use passive reflective markers and others use active optical markers (diodes). No matter the system, motion capture relies on multiple cameras being able to see each marker since, for complex movements selected points can easily become occluded by the measured object's orientation. At least two cameras must be able to see each marker at all times for accurate reconstruction of XYZ coordinate data.

Using sophisticated software, precision motion capture measurements can be assembled in 3-D computer space, supplying kinematics (positional data, distances, velocities, and accelerations) of the movement. For human subjects, markers need to identify functional anatomical points, such as joints (Lundberg,

1996). Typically, a special garment that effectively eliminates the relative movement of clothing and skin, but without restricting motion, is worn by subjects; markers are appropriately positioned on this garment. For example, upper body movement can be quantitatively characterized using markers placed on trunk and joints as follows: the Sternal end of Clavicle, Xiphoid process of Sternum, 7th cervical and 10th thorax vertibra, for the trunk; right and left Acromion, lateral Epicondyle, and Styloid processes of Radius and Ulna for the arms, and; 3rd metacarpal and finger joints for the hands. Head orientation can be determined by placing markers on the head. Frequently additional reference markers are used to determine segmental rotation. If any equipment is used during the activity, its movements can be captured by placing markers on the items. In all instances at least three markers are needed for each measured body segment and each piece of equipment in order to determine both positional translation and rotation.

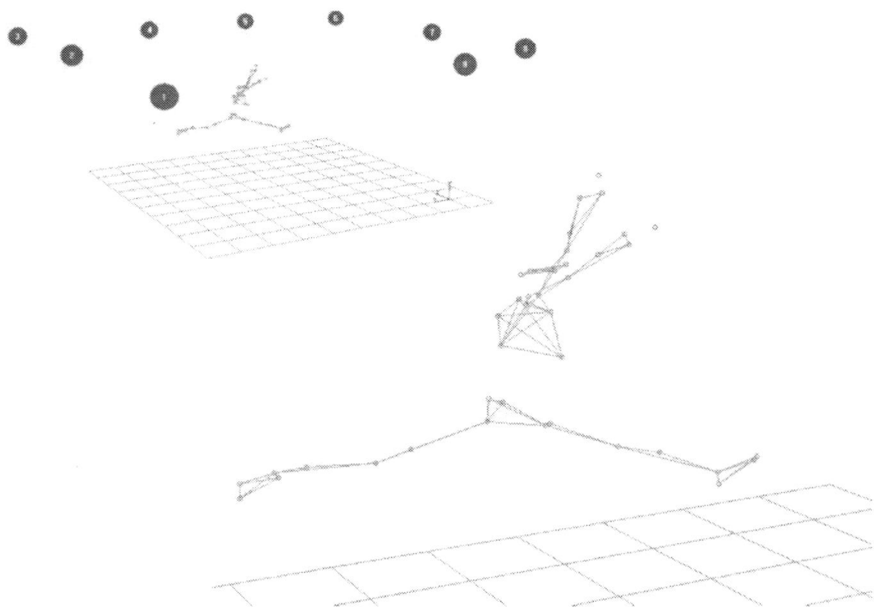

Figure 9. 3. D motion caption of Grande Jeté (Ballet): the set-up of the 9 cameras and a sample frame (G. Shan, 2005).

Biomechanical Modeling

From changes in position over time, computer analysis takes simple kinematic data and translates it into skeletal and muscular movement using the fundamental precepts of physics and anatomy. This is biomechanical modeling. In such biomechanical modeling, inertial characteristics of the body are estimated using anthropometric "norms" found through statistical studies (Shan & Bohn, 2003). Effectively, biomechanical modeling can serve as a platform for data analysis, providing the means to integrate information collected multiple measuring systems such as motion capture, force measurements and EMG.

Task-specific biomechanical models are used for each activity. Modeling techniques such as dynamic and inverse dynamic analysis allow researchers to establish joint coordination (motor control patterns), internal loads (moments and forces) and muscle working patterns (loading types). This information is used to study interactions between humans and their tools (e.g. a violinist and his bow or a golfer and her club), the interface of the body with its surroundings (e.g. dancing) muscle group timing/coordination (e.g. multi-joint tasks such as piano playing), and neural control (Wolff, 2008).

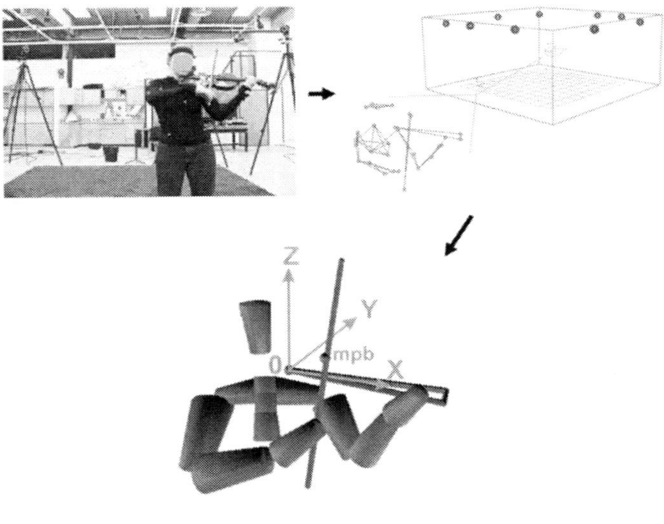

Figure 10. The set-up of 3D motion capture system and the 10-segment biomechanical model of violin playing.

For example, studies by Shan & Visentin (2003 & 2004) analyzing violin performance used 30 reflective markers to track professional violinists' left and right-arm skills (Figure 10). Cameras were placed in a circle around the subjects, creating a capture volume of approximately 3m × 3m × 2.5m. Of the thirty markers, 24 were placed on the upper body of each subject and 3 were positioned on each the violin and bow. Marker trajectories were used to reconstruct a biomechanical model of violin performance.

Electromyography

Electromyography appears to be a seductive technology because it provides researchers easy access to data about the physiological processes that cause muscles to generate force, produce movement and engage in the countless functions that allow humans to interact with their world. By placing electrodes on the surface of the skin directly above the longitudinal and medial-lateral centre of a muscle, minute electrical activity can be detected during muscle contractions. These seizure potentials can be directly correlated with the intensity and duration of muscle activity. As such, EMG supplies researchers with a means to determine the timing of neural control and to fractionate the response process of coordination among muscle groups.

EMG's main attraction is that we can, to some extent, examine what happens to muscles inside the body using non-invasive means. In absolute terms, seizure potentials measured by EMG are very small. So, an amplifier is used to enhance electrical signals, which typically register under 10 millivolts. Further, the amplitude of the signal is dependent on the physical characteristics of the subject (e.g. skin conductivity, thickness of fat layers between muscle and skin, training-dependent neural control, etc.). In actual fact, comparison of raw data among subjects is senseless. Most significantly EMG signals are stochastic (random) in nature. So, EMG raw data requires significant amounts of processing before it is of any practical use. These would seem to be significant challenges to EMGs utility but processing provides a means to overcome such barriers.

In order to interpret EMG data, which is a relative (subject dependent) rather than an absolute value (subject independent), a variety of processing protocols exist. Enveloping is the one most commonly found in the literature (Winter, 1984; Nandedkar & Sanders, 1990; Shiavi, Frigo & Pedotti, 1998). The enveloping process involves full-wave rectification of raw EMG signals and low-pass filtering (Figure 11). Such filtering improves EMG sensitivity as a diagnostic tool by keeping dominant components of the signal and eliminating unstable ones. According to the literature, a 6 Hz filter Fast Fourier Transformation (FFT) low-pass filter is typical for human activities (Arsenault, Winter & Marteniuk, 1986; Shiavi et al.1998).

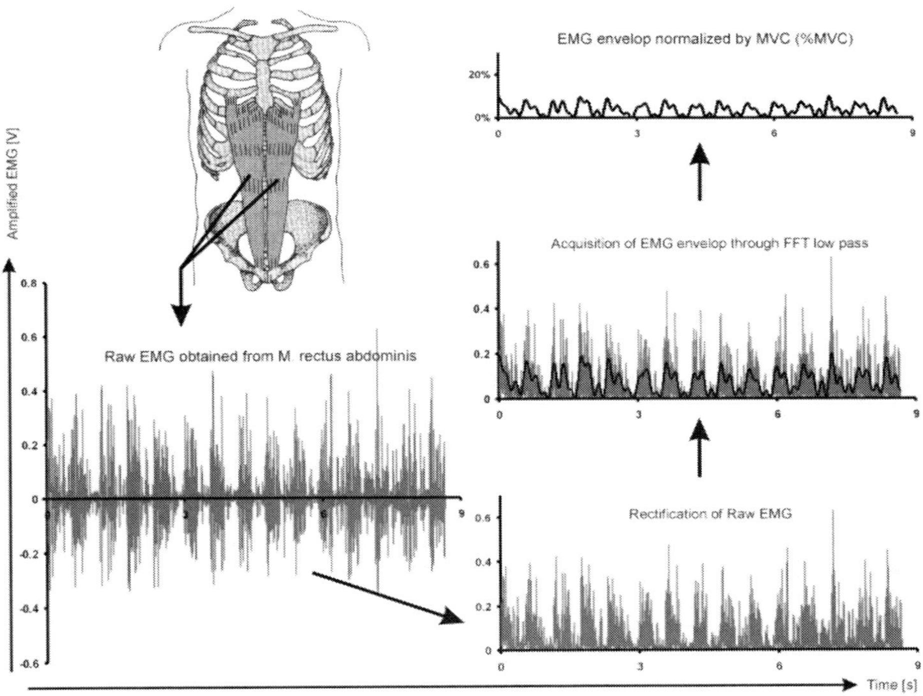

Figure 11. EMG of rectus abdominis during voluntary breathing – the process establishing enveloped and normalized EMG for quasi-quantification.

Enveloping provides a stable signal for researchers. However, it must be noted that such signals are *non-quantitative* measures of neural-musclar activity. Signal intensity cannot be used as a comparative value between subjects, or even between trials of the same subject held on different days, since it is impossible to exactly replicate placement of electrodes and other conditions that influence the quality of EMG signals. Hence in order to provide *quasi-quantification* and possibility of comparison between trials, a normalization process is required. Normalization expresses all measurements as a percentage of a chosen reference value. For EMG, commonly used references include: 1) the maximum signal level during a trial, or 2) a Maximum Voluntary Contraction (MVC) determined in a separate trial under high-load conditions. Using the first of these provides results of questionable quantitative utility since the "reference" may vary from trial to trial. The second approach establishes the physiological limits of the muscle under static contraction, a value which will not change from trial to trial. This allows researchers to examine muscle load intensity in percentage terms of the subject's MVC (or physiological limits). Such a value can provide a reliable means of comparison among different subjects and trials.

Figure 12 is a composite of three different trials measuring the rectus abdominis activity of a professional opera singer (mezzo-soprano). The rectus abdominis MVC was determined in a separate trial under high loading conditions. The protocol involved two sets of singing four short onsets (intonation of a pitch) followed by a sustained tone produced on the same pitch. There was approximately three seconds of rest between the sets. The same protocol was used at three different pitch frequencies – high (black), medium (dark grey) and low (light grey) for a typical mezzo's vocal range.

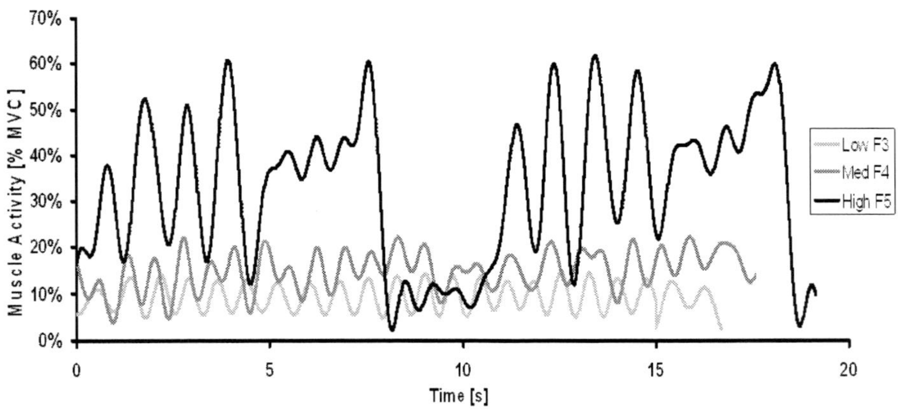

Figure 12. Rectus abdominis activity at different pitches.

In terms of vocal production, analysis of Figure 12 supplies rich details. Normalized EMG measurements are particularly revealing and the consistency/repeatability between sets provides evidence of high levels of training. Clearly, different pitch levels required significantly different amounts of physical effort from the singer. At low and medium pitch, there was almost no noticeable difference in physical effort between short onsets and long sustained tones. Interestingly, for low and medium pitch there was almost no difference in rectus abdominis activity during periods of singing or rest. Physical effort remained under 20% of the subject's MVC. At high pitch, the story changes significantly. Peak levels of EMG were above 60% of the MVC. This represents a physically demanding task. For the four short onsets, a clear pattern existed – first onset lower and remaining three higher – one that may be explained in terms of the individual's technique. Notably, the period of rest for the high frequency trial, showed lower values than the periods of rest for the other two "easier" ones (low and medium). Further, the part of the trial where a tone was sustained began with an onset of about 40% of MVC and gradually increased to peak levels of the short onsets as the breath of the singer was expended. These two facts suggest subconscious pre-planning on the part of the singer, since experience would tell her that an increase of physical effort was going to be required during the generation of a sustained tone. Such results have implications for vocal pedagogy, since singers may not be perceptually aware that such patterns are developing. Clearly, in spite of its quantitative limitations, EMG can provide significant information in terms of skill analysis and evaluation of training.

Force Measurements

Force measurements provide quantitative measures of the interaction between a human body and its surroundings. They may be determined using platforms, small sensors, arrays of sensors, or in a myriad of other ways specifically required by the conditions of testing.

In arts biomechanics, force platforms can be used to quantify ground reaction forces and weight transfer during activities such as dancing or balancing. Figure 13 shows a ballet pirouette. The yellow arrow shows the force vector of the ground reaction force in relation to the dancer's inertial centre of gravity. Not only can such information be used for pedagogical purposes, it has potential industrial significance for design of ballet footwear and for injury prevention and risk management regimes (Chockley, 2008; Shippen & May, 2008; Shippen & May, 2009; Solomon & Solomon, 2004).

Small sensors in a variety of configurations can be used to answer task-specific research questions. Examples in arts biomechanics includes installation in keyboards (Parlitz, Peschel, & Altenmuller, 1998), in mouthpieces of wind instruments (Devroop & Chesky, 2002), and on places where performers fingers or thumb come into contact with an instrument (Nemoto & Arino, 2007). The data gathered gives information about finger pain, efficiency and effectiveness of physical effort and the effects of long-term training.

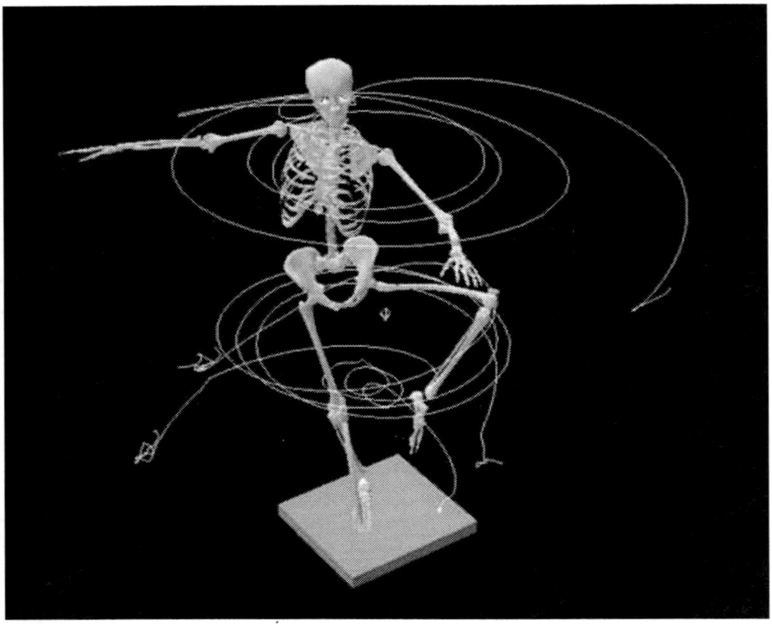

Shippen, 2009.

Figure 13. Measurement of ground reaction force using platform during dancing.

Synchronized Measurements

Together, data collection systems such as the ones described supply information on how individuals interact with their environments and tools. The data delineates cognitive and neuromuscular processes used while preparing to produce and during the execution of an action. Synchronized systems – those where data from multiple systems are measured and time-coded simultaneously – are the best means to analyze and understand the complexities of human performance because they provide multiple perspectives of the same phenomenon. The real challenge lies in find meaningful ways to integrate these various perspectives into a holistic view. From an analytic standpoint, it is problematic that inputs can include many different time variables (energy, force, linear and angular positions, speeds, accelerations, bio-signals), each of which may be measured in different units, sampling frequencies and scales. Thus, for meaningful quantitative research, a central analytical tool needs to serve as a nexus for comparing and correlating the multiple signals that are generated. Biomechanical modeling provides just this means (Figure 14).

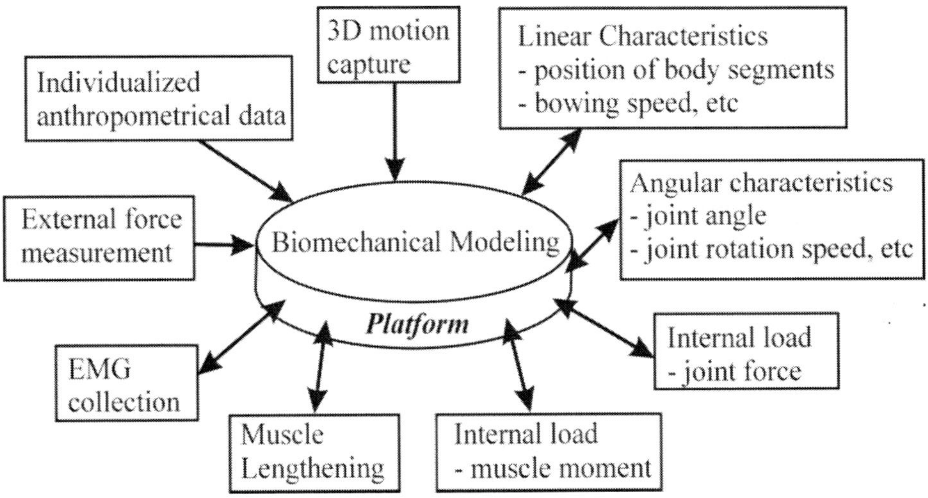

Visentin & Shan, 2004.

Figure 14. Biomechanical modeling as a platform for integration. Modeling supplies the means to connect and compare multiple inputs. Single-headed arrows represent inputs that are measured. Double arrow heads represent those that both influence *and* are influenced by the model.

In summary, synchronized systems coupled with biomechanical modelling reveal more anatomical and biomechanical perspectives than measurements alone. 3D motion capture system supplies the spatial position, velocity, and acceleration of limbs and/or joints. EMG captures muscle activity, which can be utilized to fractionate the response process and get intensity and timing of neural control as well as the coordination between and among muscle groups. Force measurements reveal the interactions between subject and environment. Video other means provide a traditional external view for motion analysis reference. Biomechanical modeling links the coordination of the joints with neural controls (muscle activities). In total, synchronized systems present a more global technique to measure, analyze and understand any human motor skills.

2.5. SAMPLES OF PEDAGOGIC APPLICATIONS

Section 2.3 began with a discussion of the principal challenges faced by Arts Biomechanics as a field of study and ended by identifying two questions that Arts Biomechanics can begin to answer. The first of these, "Can we identify, describe, and quantify qualities that lead to artistic success while minimizing the potential of performance injury?" directly addresses pedagogy. Its goal of reducing the potential of performance injuries will ultimately be unsuccessful unless it provides information for improving pedagogy. Both teachers and learners are primarily interested in much more primitive questions, such as "How can I teach/learn skills faster?" Concerns for injuries typically come low on both students' and teachers' priority lists until they have actually occurred. By engaging in research that addresses both effectiveness and efficiencies in performance, Arts Biomechanics can begin to influence attitudes. If performing artists believe there is a faster/better way to do something, they will listen. But, research in Arts Biomechanics will have to provide practical and convincing scientific evidence, clear language and definitions, and a means to re-evaluate and improve pedagogy in ways that do not compromise artists' perceptions of the end product – their art.

Unfortunately there are relatively few examples of quantitative research in the literature. One example, related to violin, gives a glimpse of the potential of Arts Biomechanics applications in pedagogy (Visentin et al., 2008). In the study the authors set out to determine whether or not it was possible to accurately and objectively characterize and generalize some biomechanical skills related to violin performance. Both smooth and bounced bowing techniques were examined at various speeds and on all four strings of the violin. The goal was to accurately identify postural & kinematic landmarks, kinetic "feel", and motor control patterns (i.e. ranges of motion of single joints and dynamic interactions among joints) that might characterize the movements of expert performers. From a pedagogical standpoint, consistencies observed among expert subjects of differing anthropometric proportions can provide "markers" that should be easily employable in teaching.

The results of the study showed that, for the two identified bowing techniques, finer motor control movements, such as those found in the wrists, were highly variable among subjects. However, shoulder and elbow movements, those joints that dominated gross motor control, provided generalizable patterns. Using these, several pedagogical markers were identified and quantitatively described. These markers provide information about posture, positions, and time-based dynamic interactions among joints during performance.

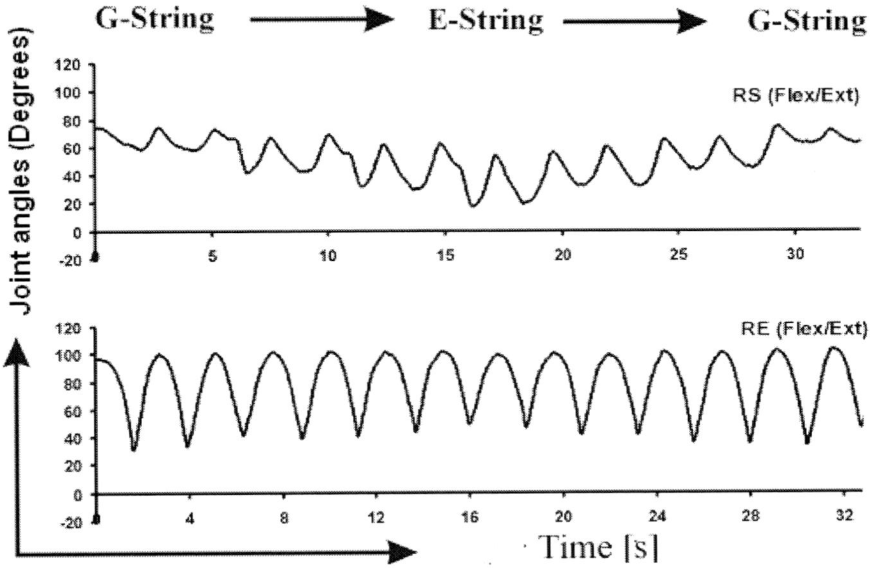

Visentin et al., 2008.

Figure 15. Right elbow and shoulder flexion/extension. Ranges of motion (ROMs) are measured by peak to trough differentials.

Figure 15 shows both that there were absolute changes in the amplitudes of the joints measured and that the control pattern changes of the two joints appeared to be complimentary – one ROM gets bigger as the other gets smaller. From G string to E string overall shoulder flexion/extension angle decreased and shoulder flexion became more active and the ROM increased. The elbow acted in exactly the opposite manner. This is one example of establishing a compound "marker" for a multi-joint task; one that can be used during the teaching/learning process.

Additionally, through biomechanical modeling the study identified tempo-dependent phases of motor control. Three discrete control pattern phases were found for both smooth and bounced bowing (G. B. Shan et al., 2007; Visentin et al., 2008). Forward dynamic analysis revealed each phase to have distinct combination of kinematic factors. The first phase was marked by increased bow acceleration and speed (indicators of exertion), while travel distance of the bow decreased (reaction showing a compensation). The second phase was characterized by constant bow acceleration, revealing workload to remain steady, while distance and speed decreased (showing efficiency gains). And the third, revealing an approach to physical limits of the player, showed bow acceleration, speed and distance to all gradually approach their respective plateaus (Figure 16).

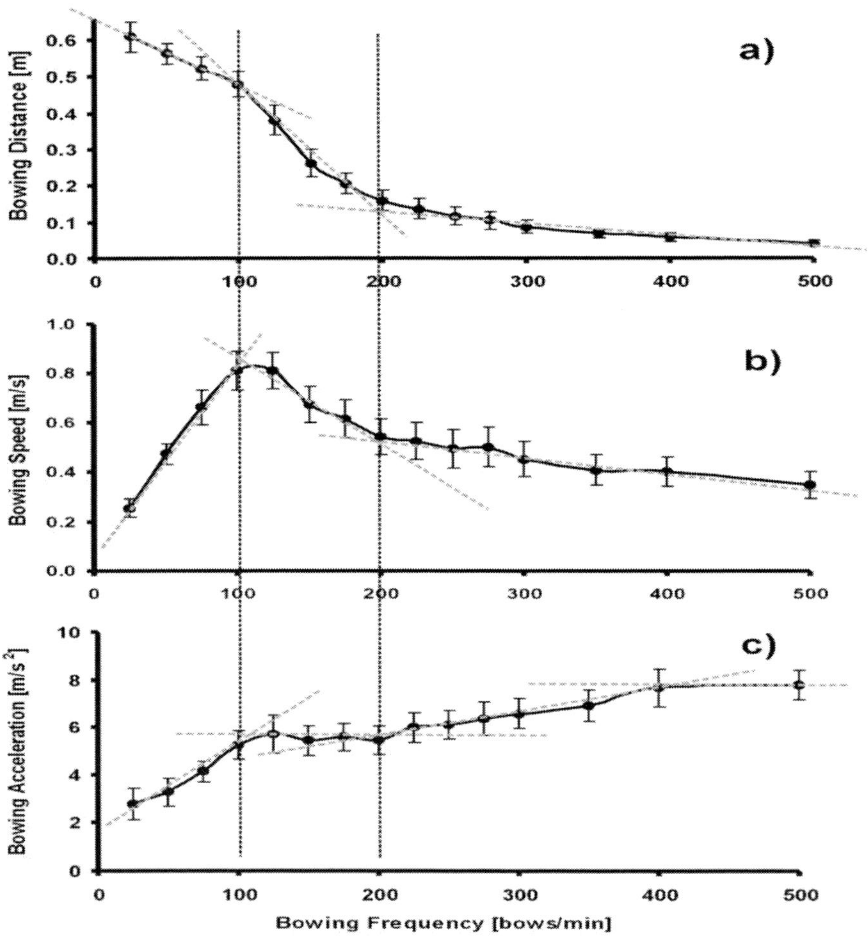

Shan & Visentin, 2003.

Figure 16. Kinematic factors showing three discrete phases of motor control defined by two critical points (vertical dotted lines). From left to right: increasing physical effort, optimization, and approaching physiological limits.

Although there are relatively few examples in the pedagogy literature, other sources contain information that also has pedagogic implications. The problem with this is that the venue of the information might preclude reaching those who are most influential in the artistic practice – performers and teachers. One such example comes from the medical literature (G. B. Shan & Visentin, 2003). The study was directed toward quantifying biomechanical factors related to Repetitive Stress Injuries (RSIs). Findings suggested that music education would benefit by applying biomechanics in the teaching and learning process. For example, Figure 17 represents a trial typically found among professional violinists. The elbow heights of the subject were normalized as a percentage of his body height. This kind of kinematic analysis shows asymmetrical control between left and right side of the body. Elbow heights were near static on the left (approximately 68% BH), while the right showed a complex dynamism. Right elbow ROMs were remarkably greater on the E-string (9.3% of body height) than on the G-string (6.9% of body height). The implications for pedagogy are significant. When teaching about bowing, teachers efforts need to convey the dynamic nature of the right elbow; its position depending on both string played and part of bow used. Further, teachers need to be aware of how the kinematic chain actually works. The mechanism that raises and lowers the elbow is actually the shoulder. Failure to clearly express such markers can slow the learning process, limit the student's potential, or be a fundamental cause of eventual injuries.

A marker-based teaching approach can provide insight into how it actually feels to play the violin. As these studies suggested, such methodology may have significant implications for development of new pedagogy models for learning performance skills on the violin. Connecting objective data with subjective perception or "feel" is one way to accelerate the motor learning process. Teachers and students alike want to reduce the time required for skill acquisition and an evidence based approach decreases the time students must spend "experimenting" to establish basic coordination control. A recent study examining university level trumpet players also alludes to the value of quantitative analyses for the development of new approaches to pedagogy (Devroop & Chesky, 2002). Clearly, more studies are needed to develop such models, not only for instrumental musicians, but also for dance and other arts.

The selected examples show Biomechanics to be ideally suited for such task. Marker-based approaches to performance pedagogy have the potential to increase the efficiency and effectiveness of skill acquisition. In this way, Arts Biomechanics is a new application of well-established ideas. These ideas can guide the formative stages of learning and influence arts professionals and teachers. But, to be successful, it needs to meet the "Art" on its own terms. To do so, Arts Biomechanics must be willing to engage in a deeper discourse that explores the relationship between motor-control patterns (the "quantitative") and the subjective aspects of the art – the desired aesthetic outcomes; thereby preserving natural variability and individuality of artistic "style".

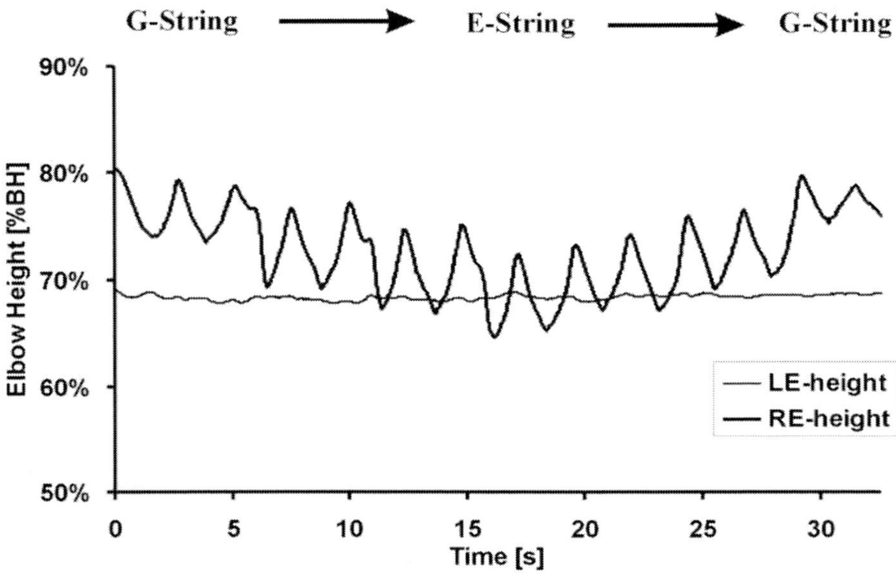

G. B. Shan & Visentin, 2003.

Figure 17. Vertical movement of left and right elbows during a typical 32-second trial. A two-octave G major scale was performed, beginning on the G-string cycling to E-string and back.

2.6. Future Directions

Arts Biomechanics is a field in its infancy. In can challenge traditional practices in arts pedagogy by successfully employing quantitative techniques to objectively describe physiological and biomechanical phenomena associated with fundamental performance skills. Systematic scientific inquiries hold the potential to provide answers to important multifaceted question such as those posed at the beginning of this section. However, arts performance is different from most day-to-day human motor activities. All arts performances are influenced by a complex set of factors related to the performers' personal aesthetic sensibilities and perceptions. Quantitative analyses can challenge these self-perceptions. The studies identified in this chapter show ways to establish a science-based approach to pedagogy. Such methodologies can increase the efficiency and effectiveness of skill acquisition, which will be welcomed by the arts industry – but only if they provide practical and convincing scientific evidence, clear definitions, and a means to reevaluate and improve pedagogy in ways that do not compromise artists' perceptions of the end product. It remains to explore the usefulness of Arts Biomechanics in the study of multiple Arts. A more comprehensive body of research could potentially revolutionize how performance skills are taught.

Chapter 3

INJURY RISK QUANTIFICATION AND PREVENTION RELATED TO ARTISTIC PERFORMANCE

3.1. OVERUSE SYNDROME

Many different terms have been used to describe the category of injuries that occur from repeated use of the human musculoskeletal system. Overuse syndrome is a "catch-all" term, familiar to the general public, for many different types of repetitive stress injuries (RSIs). Among performing artists, a myriad of injuries occur. Because artistic success or failure is often determined by micro-increments of ability, the smallest injury can negatively affect the performer and the artistic outcome. For example, an injury that affects the range of motion of someone's big toe by 10% would go unnoticed by an office clerk. However, for a prima ballerina, it could be the beginning of the end of a career. Pathologies can include: tennis elbow, shoulder injuries, rotator cuff injuries, focal dystonias, tendonitis and lower back pain and more (Fry, 1987; Zaza, 1992, 1998). Although repetitive movement may typically be the mechanism of these injuries, causal factors are deeply rooted in the Arts veneration of their traditions, pedagogic and artistic.

Reported rates of vocational injuries among performing artists are very high. Among musicians, affect between 50% and 76% of professionals (Fry, 1988; Fry et al., 1988; Zaza, 1998). Considering that this applies to musicians with injuries serious enough to affect their vocational activity, often for extended periods and sometime career-ending, these statistics are staggering. There are three general categories of injury commonly diagnosed and RSIs accounts for over half of these. Given the high incidence of RSIs, the industry has recognized the need for developing efficient and effective prevention and risk-management strategies. Currently most efforts in this regard are still in the early stages of research. Biomechanics has been contributing in the areas of accurate and quantitative description, understanding the nature of the physiological loading, and linking epidemiological studies to the observed activities. The aim has been to minimize or prevent injuries by reducing biological loading. Since Arts performance activities often involve the whole body, and demand high degrees of both fine and gross motor control, an examination of their complex biomechanical processes may lead to generalizations about RSIs that apply to other complex human activities.

Definitions influence research questions, design & structure. Therefore, it is necessary to review traditionally held definitions in order to ensure that they are sufficiently nuanced to direct researchers in the design of their studies. RSI has been traditionally defined as the excessive use of body parts so that the accumulation of micro-trauma exceeds human physiologic limits (D. M. Dawson, Hallett, & Millender, 1990). It has been understood for quite some time that basic factors influencing the etiology of RSIs include: (1) genetic makeup of indificuals; (2) performing technique; and (3) time × intensity of activity (Fry, 1987). Recent studies, identified below, improve upon these basic definitions.

3.2. Biomechanical Modeling as a Platform to Understand OS

The details of this section are drawn from an article published in the journal *Medical Problems of Performing Artists* (G. B. Shan, Visentin, & Schultz, 2004).

Injuries that occur from repetitive use significantly differ from those that typically occur from load-related accidents or during sports. In the case of the latter, injuries are essentially due to load impacts suddenly exceeding human physiological *limits* - limits that are theoretically possible to evaluate, and are relatively easy to diagnose. On the other hand, activities that cause RSIs have characteristic loads well below human physiological limits and injuries result from multiple micro-trauma, the effects of which accumulate over time. A gradual and imperceptible onset of injury shows RSIs to challenge human physiological *tolerance* - a factor that is difficult, if not impossible to quantify. Even though it may be tacitly understood by those who actually deal with such injuries, some current definitions still fail to articulate this distinction. Using the term "physiological limits" in connection with "micro-trauma accumulation" is misleading. RSIs are entirely different than impact injuries because of the time-based and cumulative nature of the pathology. Thus, for meaningful quantitative research, a central analytical tool needs to serve as a nexus for correlating the multiple signals that are generated and for situational comparison.

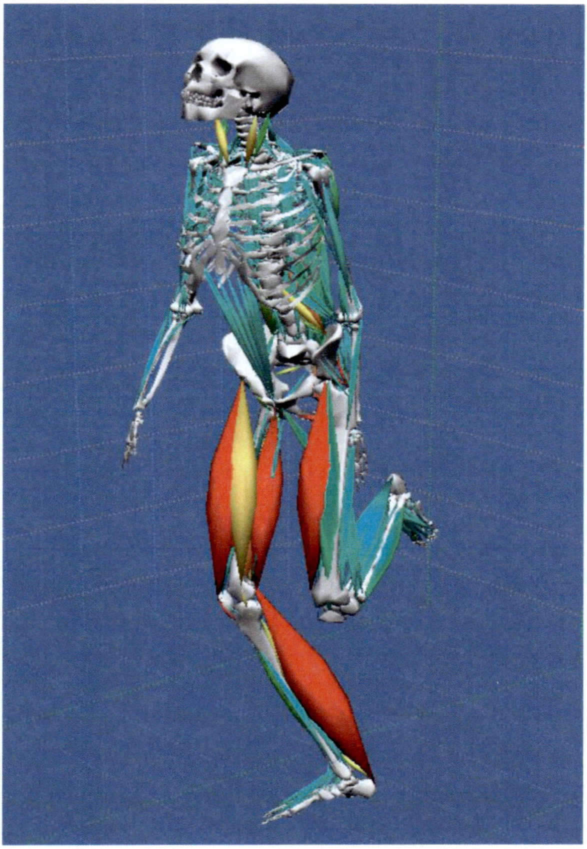

May & Shippen, 2008

Figure 18. Biomechanical model for quantification of musculo-skeletal loading and muscle lengthening/shortening.

In the case of sport injuries, biomechanical modeling has successfully served as this nexus for assessment, remediation and prevention purposes (Ballreich and Baumann, 1996; G. B. Shan, 1999). It seems the next logical step to employ biomechanical modeling as a platform in the service of RSI research. However, generalized models must first be proven to be theoretically valid for the human body, and then modified to accommodate the conditions that accompany RSIs. Ultimately, it is hoped that this will lead to research that can clearly explore the concept of physiological tolerance where: 1) a temporal factor (the process of micro-trauma accumulation) is evaluated; 2) rates of recovery during periods of relative inactivity or rest are considered; and 3) increased understanding of the evolutionary nature of OS injuries will result in changed attitudes in the workplace and a focus on the development of prevention strategies.

As it is the mechanical work performed by OS patients that obviously causes observed damage, an evaluation of "work" appears to be a logical first step. Biomechanical modeling permits quantification of both the characteristics of the human body during activity and of the workload levels (G. B. Shan and Visentin, 2003; Visentin and Shan, 2003). It is the tool best suited to this task and can be used to make connections between the data generated by multiple assessment techniques (Figure 18). Simply, biomechanical modeling allows us to understand what happens inside the body using information gathered from outside the body.

3.3. The Theory of Biomechanical Modeling as it Applies to OS

The following discussion is based on several biomechanical studies (G. B. Shan et al., 2007; Visentin et al., 2008; Visentin and Shan, 2003, 2004). Using motion capture technology, activities can be analyzed using biomechanical modeling. Anatomical positions can be obtained, the skeletal structure and simple positional data can be translated into skeletal movement using the fundamental precepts of physics. In such modeling, inertial characteristics of the body are estimated using anthropometric "norms" found through statistical studies (G. B. Shan and Bohn, 2003; Winter, 1990). Combined with motion capture information, these allow the calculation of load intensity (which can be described using a variety of parameters measured at the periphery of the body) resulting in factors such as: force, work, moment, (angular) velocity or acceleration and Head Injury Code index (HIC) (C. M. Gadd, 1966; Hennig and Lafortune, 1991; Kosiak, 1959). Biomechanical modeling allows muscles to be added onto the skeletal frame using anatomical knowledge regarding normal attachment points. Thus, muscle lengthening and shortening can be determined in connection with skeletal movement. From these, 1) internal loads (bone-to-bone contact force and muscle force, rotation caused by joint moment) can be quantified and, 2) when correlated with electromyographic signals, muscular work patterns (eccentric, concentric, isometric) can be determined.

All "modeling" involves simplifications of reality. The method's strength is that it provides a means to understand complex systems without becoming lost in detail. In the case of the human body, the key justification for using biomechanical modeling as an analytical tool hinges on the accuracy of *internal load* calculations based on the fundamental precepts of Newtonian physics (classical mechanics). To establish the contextual validity of the calculations, several steps need to be taken:

a. the various forms of mechanical work (W) as they pertain to classical mechanics must be established
b. a generalized form must be devised
c. the generalization must be justifiable and applicable to all cases for biological reactions in the human body
d. the system must be related to RSI specific conditions.

3.3.1. The Physics of Mechanical Work

Depending on the situation, work (W) may be mathematically represented as:

- Work of force F over distance d $W = F \times d$
- Work of force F at velocity v over time period t: $W = F \times v \times t$
- Work of moment M at lever arm length l: $W = M/l \times d$
- Work for accelerating (a) a mass m from 0 to velocity v: $W = \frac{1}{2} m \times v^2$
- Work for accelerating a mass m to momentum p: $W = \frac{1}{2} p^2/$
- Work for accelerating mass m by force F over time period t: $W = \frac{1}{2}(F \times t)^2/m$
- Work against friction

(c is a coefficient, n depends on the kind of friction): $W = c \times vn \times d$

Note: For the sake of simplicity, all vectors are considered constant and parallel or orthogonal to each other.

3.3.2. The Generalized Form Related to Work

In order to overcome the diversity of form observed above, it is possible to establish a generalized form covering many (or possibly all) cases through the following steps:

By appropriately substituting $d = v \times t$ and $v = a \times t$ and $p = m \times v$, these expressions can be expressed in alternate forms as follows:

$W=F\times d$	becomes	$W=F\times v\times t$
$W=F\times v\times t$	remains	$W=F\times v\times t$
$W=M/l\times d$	becomes	$W=(M/l)\times v\times t$
$W=\frac{1}{2}m\times v^2$	becomes	$W=\frac{1}{2}m\times a\times v\times t$
$W=\frac{1}{2}p^2/m$	becomes	$W=\frac{1}{2}m\times a\times v\times t$
$W=\frac{1}{2}(F\times t)^2/m$	becomes	$W=\frac{1}{2}F\times v\times t$
$W=c\times v^n\times d$	becomes	$W=c\times v^n\times v\times t$

In summary, unlike variables such as force, which cannot in this case be practically considered a parameter due to its time-dependent nature, W *may be* used as a parameter, establishing a single value for a given process. In each of the above:

- there is a case specific Variable of Intensity (VI) - which may include force, moment, stress, velocity, acceleration – plus transforming variables
- they are weighted differently by using an activity-specific exponential factor n
- loading duration t is taken into account (which may also have an exponential factor, m).

Thus, the final mathematical representation in all cases is:

$$W=C\times(VI^n\times t)^m$$

where C is a coefficient constant, n & m are integers and VI is the case-specific variable of intensity.

It should be noted that, in addition to Work, three other parameters are also frequently used for predicting biological reactions: momentum (P), load duration (t), and maximum stress (MS). For demonstrating a common structure, they can be written in unusual ways using force as a common factor. Substituting these:

$P=F^1 \times t$
$t= F^0 \times t.$
$MS= c \times \int F(t)^n dt.$

NOTE: In this last equation, the maximum force determines the value of the integral. With high n, injury occurs as $\Delta t \to 0$ and mechanical load approaches or exceeds physiological limits (Hennig & Lafortune, 1991). This is the condition of impact injuries.

From the above it can be seen that all have the structure:

$$C \times (\int F(t)^n dt)^m$$

when W, P and t are written as integrals.

3.3.3. Biomechanics and Biological Reactions in the Human Body

As the above are purely mechanical, it may be questionable that they really describe the reaction of a biological system. To demonstrate this, static work is most cited for showing the difference between mechanical and biological systems in terms of loading. When holding a mass, *mechanical work* is zero; *physiological work* is not. But this is an inadequate illustration as it confuses the techniques of "arresting" and "active holding". For example: when standing with bent knees physiologic work is needed, but when standing with straightened (arrested) knees, work is much less.

In consideration of this, empirical in-depth biomechanics studies yield parameters of biological reaction (PBRs), which are remarkably different from those yielded by mechanical ones. Examples from the literature include:

- The Gadd Severity Index $\int a^{2.5} dt = const$, predecessor of the similarly defined Head Injury Criterion HIC is widely used for predicting skull damage, where a certain constant is believed to be related to a certain degree of damage (Administration, 1996; Barth, 2000; C. Gadd, 1966; McLean & Anderson, 1997; USA, ; USGovernment).

- Re-evaluation of data indicating the p/T-limit where ulcers were seen when pressure p was exerted to the skin of dogs over time T revealed that the best PBR describing the limit is $p^{1.2}$ □ T=const. (Kosiak, 1959).
- Load which causes increase of bone density is best described by □ $\int a^n dt$, where n varies between 4 and 12, depending on the situation. (Nicol, 1998).
- World records in running are described with very high precision by $\int v^n dt$=const. The distance range studied was 12.5m to 1000 miles and could be divided into four ranges with highly constant n in each range. Between ranges, the n varied from 0.006 to 0.229. As running a world record describes the maximum load that can be bourn in this situation. This formula is a PBR too (Nicol, 1996, 1998).

In spite of the differences, it can be seen that these studies result in PBRs which have a structure in common with the mechanic ones:

$$(\int VI^n dt)^m = \text{constant}$$

The differences are that:

1. in mechanical PBRs, m may be an integer larger than zero, whereas biomechanical PBR require m=1
2. in the case of biomechanical PBRs, n does not have to be an integer because the empirical data may show a "best fit" to require a non-integer.
3. besides PBRs that can be written as integrals over the whole duration of VI, biomechanic PBRs are used which relate to only a part of the activity duration. The criterion of Hennig and Lafortune (1982) is an example. It looks only at the range between the onset and the maximum of the VI.

3.3.4. Characteristics of the PBR related to OS

As it is the mechanical work performed by OS patients that obviously causes observed damage, a work-related model of injury risk has validity. For OS the PBR can be made situation (activity) specific by examining internal and external loads and integrating their influence into risk assessment. Injury risk predictions may thus be a product of this PBR. Whereas impact injury theory examines loads that exceed physiological *limits*, OS occurs under different circumstances and requires an investigation of physiological *tolerance*.

The following simple model of OS is considered:

- OS is the result of the accumulation of micro-trauma
- activity causes micro-trauma damage over time
- biological repair takes place during periods of relative inactivity
- the cycle of 2 & 3 above is repeated multiple times
- with the result that the micro *damage* accumulates because the repair cycles are insufficient for complete recovery
- finally, a critical limit is exceeded, macroscopic injury may be diagnosed, and the *physiological tolerance* of the individual is surpassed.

In quantitative terms, the following fundamental premises may be used:

1. Since damage (D) = MicroDamage × T, it is obvious to put damage proportional to time T, because microdamage may be considered a constant that depends on the loading situation and on the properties of the biological system (both of which change minimally).
2. As repair (R) = R × T. The repair parameter R may be considered constant too. (As an alternative, R can increase or decrease as total damage changes due to the ability of the biological system to exercise repair under different conditions – i.e. the body may become more or less capable of repair at different levels of injury. Considering R to be constant represents a midpoint between these alternatives.).
3. Cumulative Damage (CD) is thus the difference between damage and repair.

$$CD = D \times T - R \times T.$$

Figure 19 shows the development of CD over a single cycle where slow repair over a long time leads to complete repair (blue lines), a single cycle where fast repair over a long time leads to complete repair (green) and repeated loading cycles (red) where CD finally exceeds the critical CD value (orange).

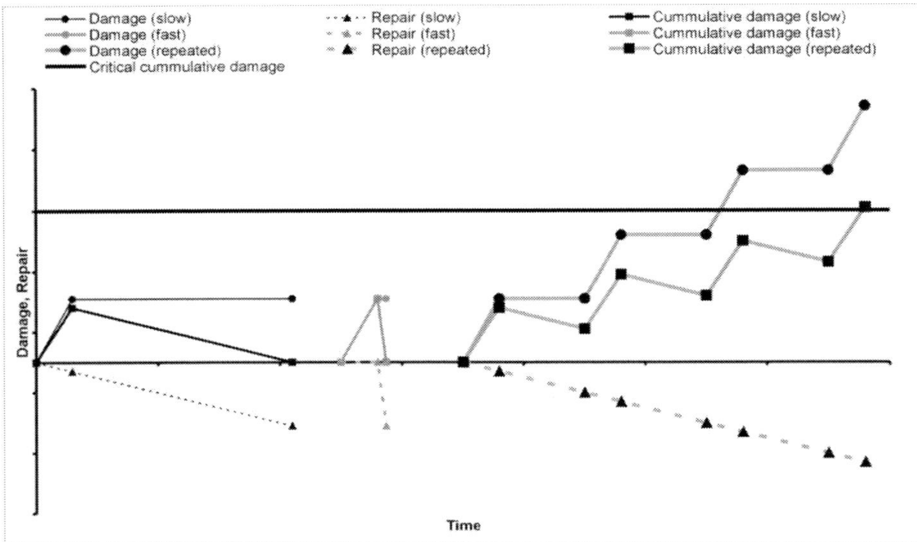

Visentin & Shan, 2004.

Figure 19. Cumulative damage scenarios from cycles of micro-damage and repair.

From this and other current studies (Visentin et al., 2008; Visentin & Shan, 2004), the following hypotheses may be made:

1. that loads leading to Overuse Syndrome (OS) have *qualitative* and *quantitative* factors, both of which must be considered when undertaking risk assessment and developing prevention strategies
2. that *qualitative* factors are related to the type of load (impact, fundamental, static, quasi-static, or dynamic) and the *quantitative* factor is related to the magnitude and duration aspects of the loading
3. that, at comparable load levels, static or quasi-static loads are more injurious than dynamic ones; making *quality* of load a primary factor to consider when assessing injury risk
4. that the intensity of a series of dynamic load cycles with small intermediate recovery phases may be considerably higher than the intensity of static load of equal duration without causing injury
5. that risk is directly related to the PBR and Intensity^n × Duration of Activity, and
6. that Intensity and n are influenced by *quality* and *quantity* of load.

In quantitative time-dependent terms, the PBR is directly related to activity Intensity (I), which is contingent on the patterns of the load for the activity in question. Therefore:

$$PBR \propto \int (I)^n dt.$$

For RSIs, the *intensity* of activity is very low, work patterns change over time, and duration is the prime factor. Such conditions make RSI research extremely complex since establishing injury risk requires that: 1) intensity (I) be measured for the activity; 2) duration of activity be integrated into the risk calculation, resulting in a context-specific PBR; and 3) that rates of repair during periods of rest be considered. Unlike impact injuries where there is virtually no time between loading and injury ($\Delta t \rightarrow 0$), repair cycles during periods of relative inactivity may not be neglected in the case of OS. In representing this model mathematically, the repair parameter (R) must be considered. Thus, Accumulated Micro-Trauma (AMT) is the net of these two biological processes:

$$AMT = \int (I)^n dt - \int R dt$$

All research must be based on some fundamental theory. This work establishes a more nuanced general form to guide RSI research in the future.

3.4. Current OS Studies Related to Music Performance

Systematic research into artists' vocational injuries is relatively young. The primary venue for knowledge transfer has been the journal *Medical Problems of Performing Artists*. It has a 27-year history examining epidemiology and remedial treatments for artists' injuries. It has long been acknowledged that prevention was an important, but under-addressed area. To promote research in prevention, the journal has been very accepting of cross-disciplinary research that provides vocational views and/or points in the direction of remediation. Quantitative biomechanics studies have been an important part of providing linkages between epidemiology, remediation and prevention since 2003. The earliest of these studies examines biomechanics of violin performance (G. B. Shan & Visentin, 2003; Visentin & Shan, 2003). Prior to this, some qualitative biomechanics work was included and, although it implicitly showed the potential to influence prevention, the results of the studies were essentially just descriptive (Tulchinsky, 1995; Tulchinsky & Riolo, 1994b). One of the missing elements was a theory or framework guiding future steps in the process.

Richard Dawson, one of the founders of *Medical Problems of Performing Artists,* has published two retrospectives documenting the kinds and numbers of studies found in the international literature pertaining specifically to artists' injuries (W. J. Dawson, 2003, 2007). These provide insight into the state of research involving biomechanics and the Arts. Of 2385 studies identified by Dawson (between 1997 & 2006), only 23 involved biomechanics. The vast majority of these used the techniques and technologies of biomechanics to examine causal factors leading to repetitive stress injuries. Below are some examples of Arts Biomechanics research in the literature:

The body of work by Shan and Visentin has been formative in establishing a sequential series of steps and a fundamental theory underpinning the utility of Arts Biomechanics in RSI research (G. B. Shan & Visentin, 2003; G. B. Shan et al., 2004; G. B. Shan et al., 2007; Visentin et al., 2008; Visentin & Shan, 2003, 2004). From phenomenological description to multidimensional signal analysis, Arts Biomechanics was shown to hold significant potential in the areas of loading quantification and injury risk management. Equally important, it was shown to have potential in the areas of changing vocational attitudes and pedagogy. One of these studies identified three factors influencing the relationship between muscle loading and RSIs: load quantity (intensity), quality (static versus dynamic), and duration (time) (Kuipers, 1998; Taylor, 1996; Visentin & Shan, 2004). Regarding load quantity, the study showed muscle activity during violin performance to be within the "safe" range as typically identified by the sports biomechanics literature (Peterson & Renström, 1987). Given the frequency of injury among violinists this is obviously not the case. Shan and Visentin's work also considered load quality and duration. Using multidimensional signal analysis, loading patterns (qualities) in the muscles and joints were correlated with both load quantities and injury rates. Because violin performance is an asymmetrical activity, this provided a means of comparison between static and dynamic loading. Results had implications for the relative risks of static vs dynamic loading and began to quantify relative levels of risk. They also pointed toward pedagogical strategies aimed at prevention. Generally speaking, since static loading is more dangerous than dynamic (for the development of RSIs), the study suggested minimizing static conditions or postures, and encouraging dynamic motor control. Further the body of work addressed the issue of disconnects between performers bodily self-awareness during performance and the quantitative reality of the phenomenon. For example, violinists are typically taught that the left elbow should have a degree of lateral movement to accommodate playing on different strings. This improves performance outcomes in the areas of tone production and pitch tuning. Unexpectedly, they found that performers generally locked their shoulders in a highly adducted position; a posture that creates static conditions and increases the risk of injury to the left shoulder (Visentin & Shan, 2003).

In another violin study by Rabuffetti et al. (2007), the effects of using different violin shoulder rests during performance were quantitatively described. The positions of markers placed on the player's body, violin and bow, permitted a kinematic analysis of position changes resulting from the use of different shoulder rests. The study identified that, since the instrument has a fixed geometry, the performer adapts to the instrument throughout the learning process. Shoulder rests are one of the few tools available to adapt the instrument to the performer. Rabuffetti et al concluded that "optimization of the violin setup cannot be based on objective evidence only (anthropometric size, violin features, type of repertoire, etc.) but still requires a trial-and-error approach." Motion analysis was used as means to provide information that might accelerate this process (Rabuffetti, Converti, Boccardi, & Ferrarin, 2007).

In a study of trumpet mouthpieces, Devroop and Chesky (2002) examined intraoralpressures, and compression of the mouthpiece against the embouchure. The basic assumption was that these physical elements contribute to performance related medical problems among trumpeters. Results reinforced concerns regarding risk for performance-related medical problems, including rupture of the orbicularis oris muscle, tooth displacement, head and neck lesions, embouchure dystonia, and temporomandibular joint (TMJ) and jaw pain. In spite of the fact that many pedagogues and performers had attempted to explain some of the mysteries surrounding mouthpiece forces the study contended that their relationship was still not fully understood because these explanations were based upon empirical evidence and not quantitative measurements. The study confirmed that laboratory-based research can be effective in searching for answers to real-world problems of performing artists (Devroop & Chesky, 2002).

In a 3-dimensional biomechanics study of pianists' hand and finger movement patterns, Virgilio et al. (2007) analyzed the movements recorded using a multi-camera motion capture system. A complex musical sequence, 16 measures of a Bach Minuet, was used as a test protocol for estimating the actual load imposed on joints during playing, as well as to assess the influence of experience on the actual movements. The central hypotheses blended concepts of injury prevention and pedagogy – that different patterns of movement would be used depending on the player's experience and that pianists could use different techniques to reach the same results. The study identifies that: 1) the movements required for piano playing usually involve low impact loads that do not exceed physiologic limits of human body, but their repetition may provoke microtrauma leading to overuse injuries; 2) Experience may allow a pianist to modify the motor patterns used for a performance, allowing the highest accuracy with minimum effort; and 3) that "didactic techniques should always teach the students how to reduce—or better, to avoid—professional disorders" and "the use of a wrong technique may accelerate the development of overuse stress injuries". Such work underpins the premise that prevention is a matter of risk management during the teaching and learning process (Virgilio, Chiara, Emilia, Paolo, & Sforza, 2007).

Together, these and the other available few examples all directly or indirectly suggest a need for more research in the area of optimizing practice time and pedagogy with regard to both efficiency (skill acquisition time) and effectiveness (learning while managing injury risk).

3.5. Challenges Facing Injury Prevention Research

From the above discussion, one might observe that injury prevention, in many ways, proves to be a synonym for behavior modification. All suggest changes in working patterns or learning precesses. Indirectly, these result in trade-offs. For example, when walking on an icy sidewalk in slippery shoes, it would be wise to shorten one's stride length in order to avoid falling. This kind of control pattern change is a consciously invoked behavior modification. Of course, our fictitious walker will perceive this prevention strategy as slowing her down. If our walker is in a hurry, she faces a choice; either she may either increase stride length or frequency, balancing the risk of falling with the potential time gains she may achieve, or she must accept arriving at her destination later than anticipated. This analogy shares many similarities with aspects of injury prevention in the performing arts; carrer progress and injury prevention often seem at temporal odds.

In the highly competitive field of arts performance, the risk of "slowing down" can easily be perceived as more vocationally risky than is the potential of injury. Long hours of repeititve and demanding physical activity associated with the creation of the "art" is perceived as a prerequisite of success. In terms of performance, long-term health and wellness considerations sometimes appear to be at odds with short-term artistic aims. Learners are typically indoctrinated with a belief system that is competition driven. Amount of time spent practicing is often the first measure used to evaluate dedication and artistic potential. This pedagogical approach and its associated psychology need to be understood as drivers of artists' decision-making processes. In such a reality, researchers, educators and health and wellness practitioners face significant challenges.

The problem can be best understood by drawing from a line of inquiry from the field of philosophy – Prisoner's Dilemas. The following discussion is mainly based on an in-press article in the journal *Medical Problems of Performing Artists* (Shan & Visentin, 2010).

What is a prisoner's dilemma?

In philosophy, a prisoner's dilemma (PD) is a game-theory means to illustrate possible conflicts between individual and communal rationality. Typical PDs use constructed narratives where two individuals each have two behavioral choices, and symmetrical consequential payoffs exist. For example, two thieves suspected of having robbed a bank are questioned separately by the police. Each thief has a choice to remain silent (cooperate with her partner) or to confess (defect on her partner). As part of the scenario, there are four realistic outcomes possible (P1, P2, P3, P4), with varying consequences: 1) if both thieves cooperate, each receives a 5-year sentence; 2) if both defect, each receives a 10-year sentence; and 3) if one defects and the other cooperates, the defector gets off free while the cooperator receives a 25-year sentence. This forms a 2×2 payoff matrix such that P1<P2<P3<P4 (0<5<10<25 years) for each individual, but their collective interests may conflict (Table 1).

		Thief 2	
		Cooperate	**Defect**
Thief 1	**Cooperate**	*5, 5*	*25, 0*
	Defect	*0, 25*	*10, 10*

Table 1. Payoffs of a typical prisoner's dilemma

The dilemma is simple. Suppose Thief 1 cooperates. Her payoff will be either 5 years or 25 years. Thief 2's payoff will be either 5 years for cooperating or freedom for defecting. On the other hand, if Thief 1 defects her payoff will be either freedom or 10 years while thief 2's payoff will be 25 years for cooperating or 10 years defecting. So, from Thief 2's standpoint, no matter what Thief 1 does, defection is always the most rational individual behavior since it minimizes Thief 2's possible jail time. Because the payoff matrix is symmetrical, the same is true if the thieves' roles are reversed. This simple PD demonstrates the conflict between individual and communal rationality and decision-making. Clearly from a communal standpoint it would be best if both thieves cooperated, minimizing their total jail time and maybe even keeping their ill-gotten gains.

The Prisoner's Dilemma for performing artists

PDs can also be used to demonstrate problems within individual decision-making. These are pertinent to the case of performing artists because, typically, they show a choice/reward decision-making process forming a matrix with differing payoffs. An example by Quinn (1990) illustrates this: a medical device enables electric current to be applied to a patient's body in gradations so small that there is no perceivable difference between adjacent settings. A subject is attached to the device and given the choice every day to increase the electric current by one setting and be rewarded, or leave it where it is and receive nothing. Since there is no perceptible difference between adjacent settings, it appears rational to advance the setting each day. But after a few years the pain is so great that any rational person would sacrifice everything to be free of the machine. The worst part of Qunn's scenario is that, no matter what the current setting (e.g, zero or 1,000), since incremental discomfort is imperceptible, turning the dial one more time always appears to be a rational choice. When recounted in this manner, Quinn's PD and the circumstances leading to the development of musculoskeletal injuries in performing artists have remarkable similarities. Often, the long term consequences of the behavior result in debilitating injury (e.g. repetitive stress injuries – RSIs) yet the day-to-day perception is one of gained artistic benefit (Shan & Visentin, 2003; Visentin & Shan, 2004). This is a significant situational problem facing performing arts health and wellness educators and practitioners.

Researchers, educators and health and wellness practitioners face need to understand the reasons for an artist's behaviors and encourage activities that require the artist to cooperate in fostering his/her own long term vocational well-being. As researchers, scientists and doctors need to engage artists in a process whereby learning efficiency and effectiveness become central to both pedagogy and practice. As well, they need to be undaunted and engage in discussions of artistry. Otherwise they risk their credibility by failing to show sufficient empathy with artists' underlying motivations. Injury prevention will require artists to "buy-in" to concepts that will require them to modify their practices. Buy-in will only occur if research results can be made to appear to enhance the artistic process and improve potential outcomes.

3.6. Future Directions

Based on the current state of Arts Biomechanics research, there remains a considerable amount of work to do. Bridging the gap between quantitative research and empirical experience – thus providing vocationally meaningful research results – requires a marriage between the quantitative powers of science and the empirical/vocational strengths of the performing Arts. Multidisciplinary collaborations hold the potential to do exactly that. Integrative multidisciplinary research can provide outcomes that have practical application by co-opting the involvement and efforts of performers, teachers, medical practitioners, and biomechanics researchers. Even though term "multidisciplinary" has become catch-phrase of some currency, multidisciplinary research is considerably more difficult to do than might initially be expected. Perhaps this accounts for the dearth of research in the area. In Arts Biomechanics, multidisciplinary research holds the potential to explore the liminal spaces among sciences, fine arts and education. Because researchers need to work directly with practitioners, account for perceptual experience at all stages of the work, and engage in dialogue about disciplinary perspectives throughout the process, it is a challenging but rewarding endeavour.

Arts Biomechanics has a particularly important role to play in the area of injury prevention, risk management and remediation. The authors see this as a three stage process. *Context* is established through integrating the multiple perspectives provided by MACRO studies (kinematic, kinetic, and EMG research). Such methods reveal quantitative "external" (motor control pattern and intensity of load) parameters and qualitative internal ones (normalized EMG). They provide the first stage of a holistic approach to establish the true *nature* of biological loading. Additional explorations of OS must include the examination of context-relevant accumulations of micro-trauma on the MICRO level. This second stage of biomechanics research involves applying MACRO conditions in MICRO-level research. We speculate that existing technology can be used to determine the *process* of muscle fibre damage and patterns of activity-specific muscular trauma. The third stage must examine muscular regeneration under different circumstances. Recovery during stages of relative inactivity (rest) is a vital factor for risk management.

Stage 1 – Determination of Realistic Loading Conditions

3-D motion capture, biomechanical modeling, and EMG provide multiple viewpoints for researchers and medical practitioners. Multi-dimensional quantitative analysis integrates their several perspectives and allows a systematic evaluation of biological loading. The main difficulties in this stage come from the high number and heterogeneity of signal components in terms of sampling frequencies, units of measure, diversity of generally accepted practice in dealing with the varied raw data, and interpretation of the results. Individually and combinatorially, using several measurement techniques permits quantitative description of factors related to the mechanism of injury. Kinematic modeling confirms group effects among muscles through positional description (an external perspective) and muscle length changes (an internal view). Dynamic and inverse dynamic modeling based on kinematic characteristics establish *quantity* of load, and quantify some factors related to *quality* (impact, fundamental, static, dynamic). EMG determines the presence and general level of loading (e.g. relative muscle activity) with the normalization process pointing toward the physiological limits of the individual. Muscle lengths in combination with EMG and kinetic results create a clearer picture of the working pattern of the muscles. The main task in this phase is to measure diverse Arts performance contexts.

Stage 2 – MICRO Examination of Muscle Fibre under Realistic Conditions from Stage 1

Imaging techniques such as LM, TEM, and MRI are frequently used to examine already injured muscle areas but not the actual process of injury (Brandser, Elkhoury, Kathol, Callaghan, & Tearse, 1995; Komulainen, Takala, Kuipers, & Hesselink, 1998; Rubin et al., 1995; Uchiyama, Tamaki, & Fukuda, 2001). Traditional microscopic techniques have long been used to categorize morphological changes in stressed muscle tissue. Until two decades ago, the modalities available to *image* injuries for muscle and muscular attachments were plain radiography and to a lesser extent ultrasound. Plain radiography is ineffective because of its limited tissue contrast (Sundar & Carty, 1994). Ultrasound is highly operator dependent and has a limited view of soft tissue. More recent developments include computed tomography that can only obtain images in the axial plane and magnetic resonance imaging (MRI). In particular, MRI has recently become the modality of choice for imaging muscles, tendons and ligaments.

Muscle research using animal models shows that injuries can be detected using histological and MRI methods (Biral et al., 2000; Kanbe et al., 1998; Smith, Plyley, Rodgers, & McKee, 1997; Thompson, Balog, Fitts, & Riley, 1999). These studies focus on impact (or near-impact) loading, static loads at high levels of intensity, eccentric (lengthening) conditions, and end-state examinations of the muscles (Koskinen et al., 2002; Stauber & Willems, 2002; Vijayan, Thompson, Norenberg, Fitts, & Riley, 2001). Such methods do not mimic the conditions that lead to OS, where loads are normally far below "impact" levels, load patterns vary over time, and injuries are the result of micro-trauma accumulation. We suggest that, using realistic loading conditions determined in MACRO research, MRI techniques could be used to track the accumulation of micro-trauma using in-vivo methods, examining muscle deterioration over longer loading durations.

Stage 3 – Determination of Accumulated Micro-Trauma

Based on the current success of MRI in MICRO-level imaging of existing injuries (Rubin et al., 1995; Steinbach, Fleckenstein, & Mink, 1994), a time course series could be used to determine the effect of repetitive stimulation on a muscle, and the *rate* at which micro-trauma accumulates. Micro-level studies may also be able to explore rates of repair during cycles of rest. In Section 3.3 (The Theory of Biomechanical Modeling as It Applies to OS) OS was mathematically represented time-dependent terms as follows:

$$AMT = \int (I)^n dt - \int R dt$$

Since both of these factors involve internal biological processes, research must focus on determining activity-specific values for I and n in the first of these integrals and on R for the second.

Prevention first requires meaningful measures and then comparison to the real world. Arts Biomechanics permits the integration of information from multiple inputs with a quantitative, time-based evaluation of load. It provides linking perspectives between the process and the endstate of OS and is the only method currently available that can do so. It remains to fully apply Arts Biomechanics in the service of determining factors associated with OS.

Chapter 4

Applications of Movement Science Technologies in the Creation of Art

4.1. Overview

Movement science as a field can sometimes contribute to the arts in unexpected ways. Because innovation is a driving force in both science and art, technology can create conceptual bridges between Art and Science. In the previous chapters of this book, Arts Biomechanics was seen to provide a means to characterize intersections between body movement and artistic creation. Technology provided analytic tools and biomechanics provided a means to understand and interpret the data. This relationship can be extended to the creation of art, whether it involves actual movement or provides an analytic perspective of existing artworks. Technology frequently carries potential beyond its intended applications, but artists, composers, performers, and scientists need to work together in order to harness its latent possibilities in the service of the creation of art. Ultimately audiences will be the judge of an artwork's success or failure. Technology's role needs to be carefully considered so that in does not merely serve as a layer of complexity, confounding communication between artist and audience. Rather, the aim of technology should be to enhance the intensity and relevance of the artistic concept. The purpose of this section is to provide discussion of several such links.

In the visual arts, creative motivation typically derives from artists' emphasis of their own individuality and their subjective interpretations of phenomena. As such, the end product can be understood as an effort to represent different aspects of the human experience, as *they* perceive it. This, in fact, is not just an artistic concept, it is a scientific one, too. Both science and art are contingent upon having a means of ordering one's perception of the world and of expressing that perception. Science does so with numbers and formulae while the visual arts does so with images and sculpture. Using the approaches of science, movement analysis technology can provide a link between artistic representations of human movement and our experiential understanding of the bodily condition. The following discussion tracing some of the influence of movement science on the visual arts is based on an in-press manuscript by the authors (Shan, Visentin and Harnett, 2010).

Historically, representing time-based event of human movement in an image or sculpture has been a challenge to artists. Pre-photography, human movement was typically represented by capturing a single critical point during the unfolding of a gesture. Such stop-action representation often exaggerated the phenomenon in order to emphasize an "affect" that would elicit empathy in the viewer; such art typically relies on the viewer's experiential knowledge (of their own bodily condition) or on iconography and/or cultural loading.

> "What work is there which is as distorted and elaborate as that *Diskobolos* of Myron? But if anyone should criticize this work because it was not sufficiently upright, would he not reveal a lack of understanding of the art, in which the most praiseworthy quality is this very novelty and difficulty?" *Lucian III: The Lover of Lies* (1916) translated by A. M. Harmon (Webpage of the University of Chicago, 2006).

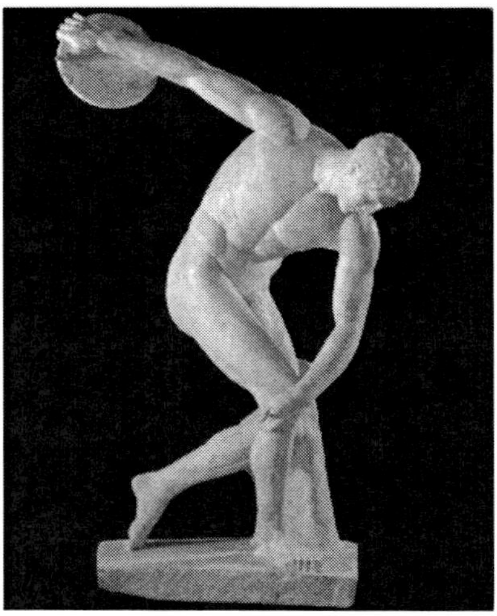

Figure 20. Diskobolos of Myron (circa. 460 BC).

For example, Myron's classical representation of a discus thrower (circa. 460 BC) is, from an anthropokinetical standpoint, hardly realistic. Yet, its representation of athleticism and dynamism are understood; the contorted, unnatural body position as a means to show power and effort has become iconographic (Figure 20).

Until the 20th century, affective representations of human movement are the norm. Developments in technology – the advent of photography – inspire some of the most profound changes in 2-D image-making. Photographic techniques were employed by both scientists and artists. The work of Edward James Muybridge may have been scientific investigations, but it had an transformative effect on the art world. Muybridge used multi-camera stop action photography (Figure 21 & 22) and multiple exposure techniques (Figure 23) to initiate a scientific approach to human movement studies (Cavanagh, 1990; Solnit, 2003).

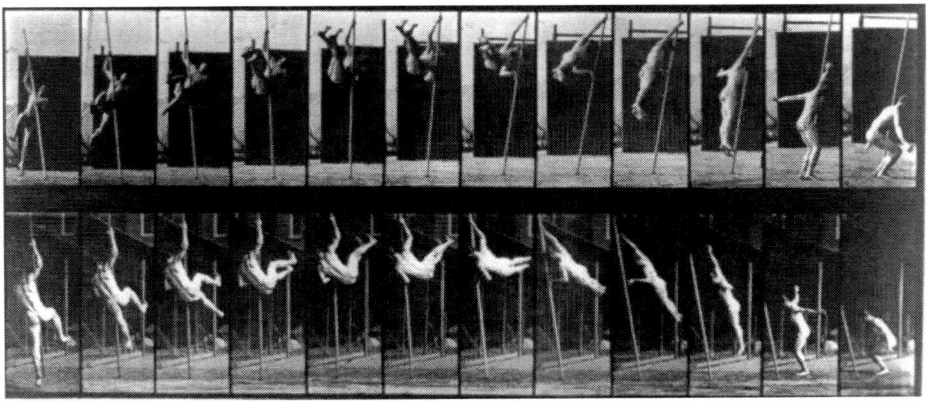

Figure 21. Multi-camera stop action recording of a pole-vaulter, by Muybridge.

Further developments using photography to capture and analyze human movement can be found in the work of: 1) Etienne-Jules Marey, who worked with high-speed photography at the same time as Muybridge, capturing multiple images with one camera on one photo plate (Braun, 1992); 2) Wilhelm Braune and Otto Fischer, who created elaborate 3-D models and performed thorough gait analyses using 4 cameras and subjects marked with Geissler tubes which emitted current that could then be seen on film (Cavanagh, 1990); 3) Nikolaj Bernstein, who developed precise kinematic procedures using cyclography, a technique employing film that was slowly moving through a camera with a mirror placed at a 45 degree angle to the optical axis, thereby creating a picture that gave a frontal and lateral view of the subject (Medved, 2001); and others – using techniques such as strobes, multiple and time exposure (Plagenhoef, 1971). Biomechanical analysis using modern-day, motion capture technology and can be considered a direct outgrowth of this work.

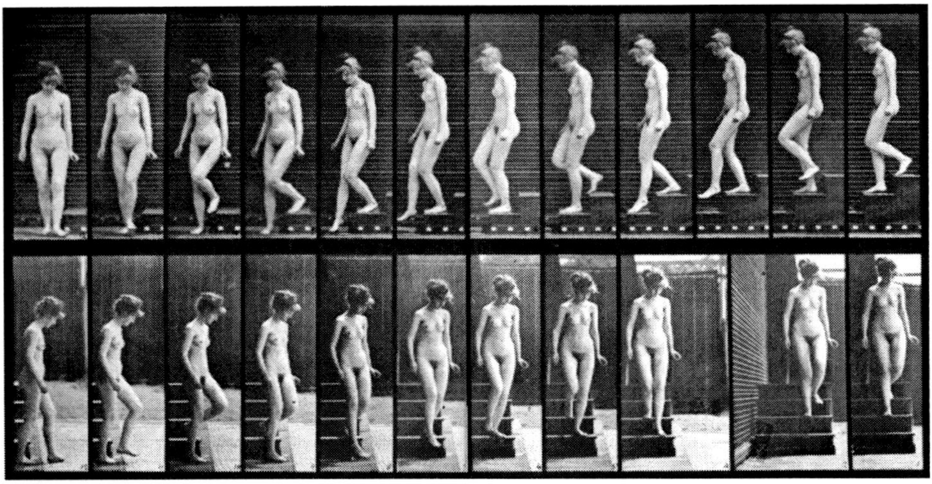

Figure 22. Multi-camera stop action of a nude descending a staircase, by Muybridge.

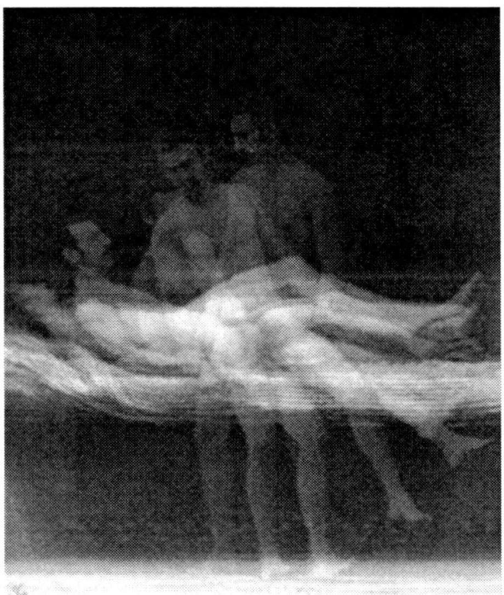

Figure 23. Multiple exposure photography documenting human movement, by Muybridge.

Directly influenced by photography, early 20th-century Futurist painters such as Umberto Boccioni and Gino Severini began to experiment with these ideas, creating works (Figure 24, *Dynamism of a Dancer [1912]*) that compressed the principle of multiple exposure imagery into "kaleidoscope art of force lines, vectored shapes and mobile patterns" (Hunter, Jacobus, & Wheeler, 2004). Displaying multiple fragmented essences in a single visual image, these artworks approach an expression of time and position designed to represent a perception of the human experience, by combining the traditional idea of stop-action affective representation with Muybridge's multiple exposure techniques.

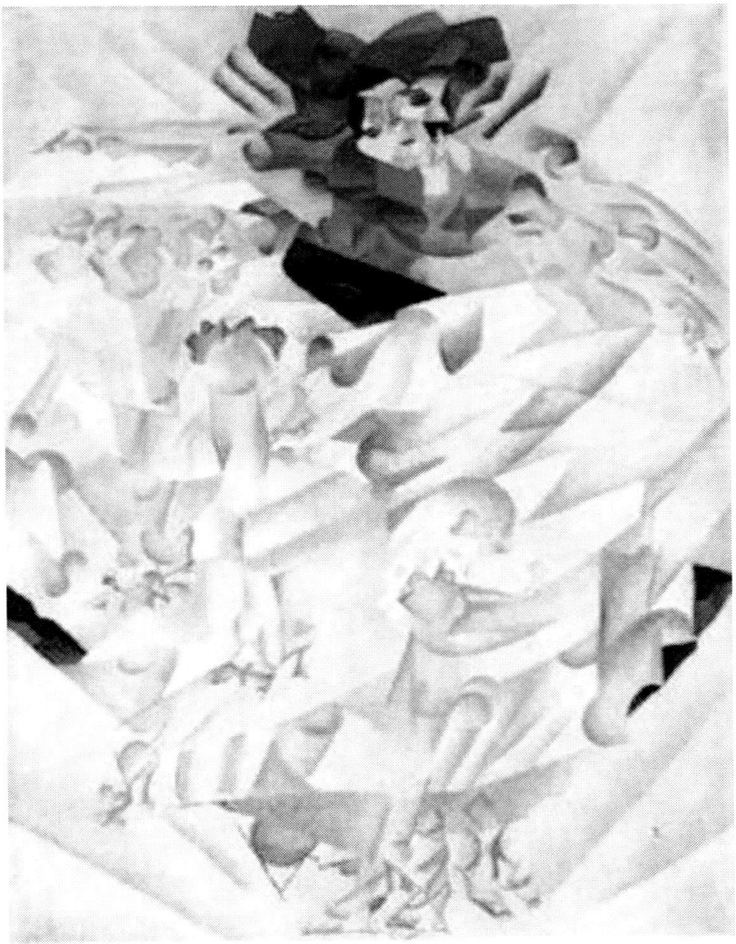

Figure 24. Gino Severini - Painting: Dynamism of a Dancer [1912] .

In each of these instances, the artwork can be understood to capture a sequential expression of how we experience existence; one where technology has influenced our existential perception of the world. American Modernist art sometimes strips down the imagery to its bare essentials, leaving only gesture. For example, Franz Kline (Kline, 1950) condenses gesture into a single frame of reference (one that may not be anthropokinetically accurate, but is bodily expressive), and contemporary artist Tanya Harnett builds the dynamics of a 3-D space through layered subject matter, gesture, and historical reference, using them as a stage for a photographic image that compresses time, x-y-z dimensionality and cultural reference into a flat-surface gestural work. Harnett's *Nude Descending a Staircase* (Figures 25 and 26) is a direct reference to Marcel Duchamp's work of the same title, who acknowledged the influence of Etienne-Jules Marey's motion capture studies. When such works are viewed, they can be understood as a metaphors and commentaries on the human experience.

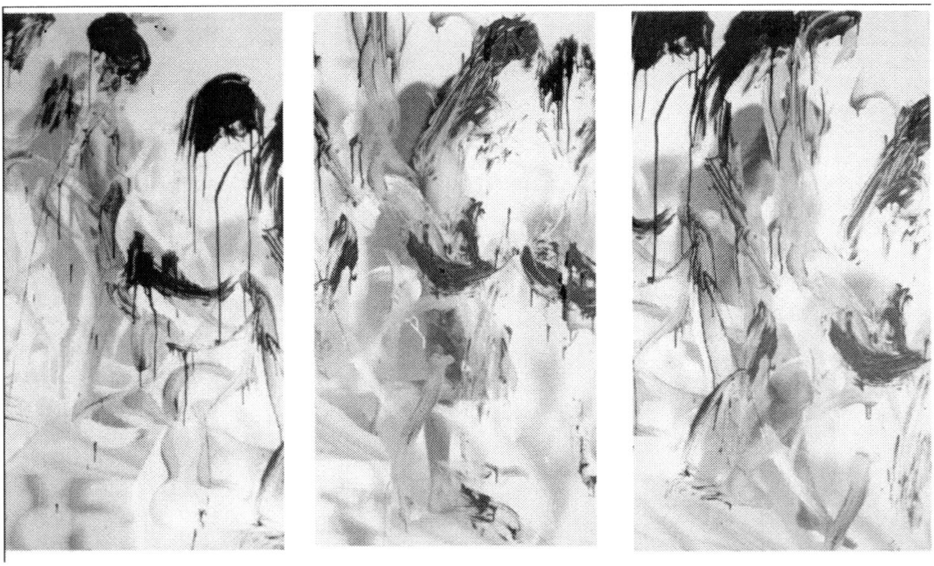

Tanya Harnett, 2008, reprinted with permission of the artist .

Figure 25. Nude Descending a Staircase.

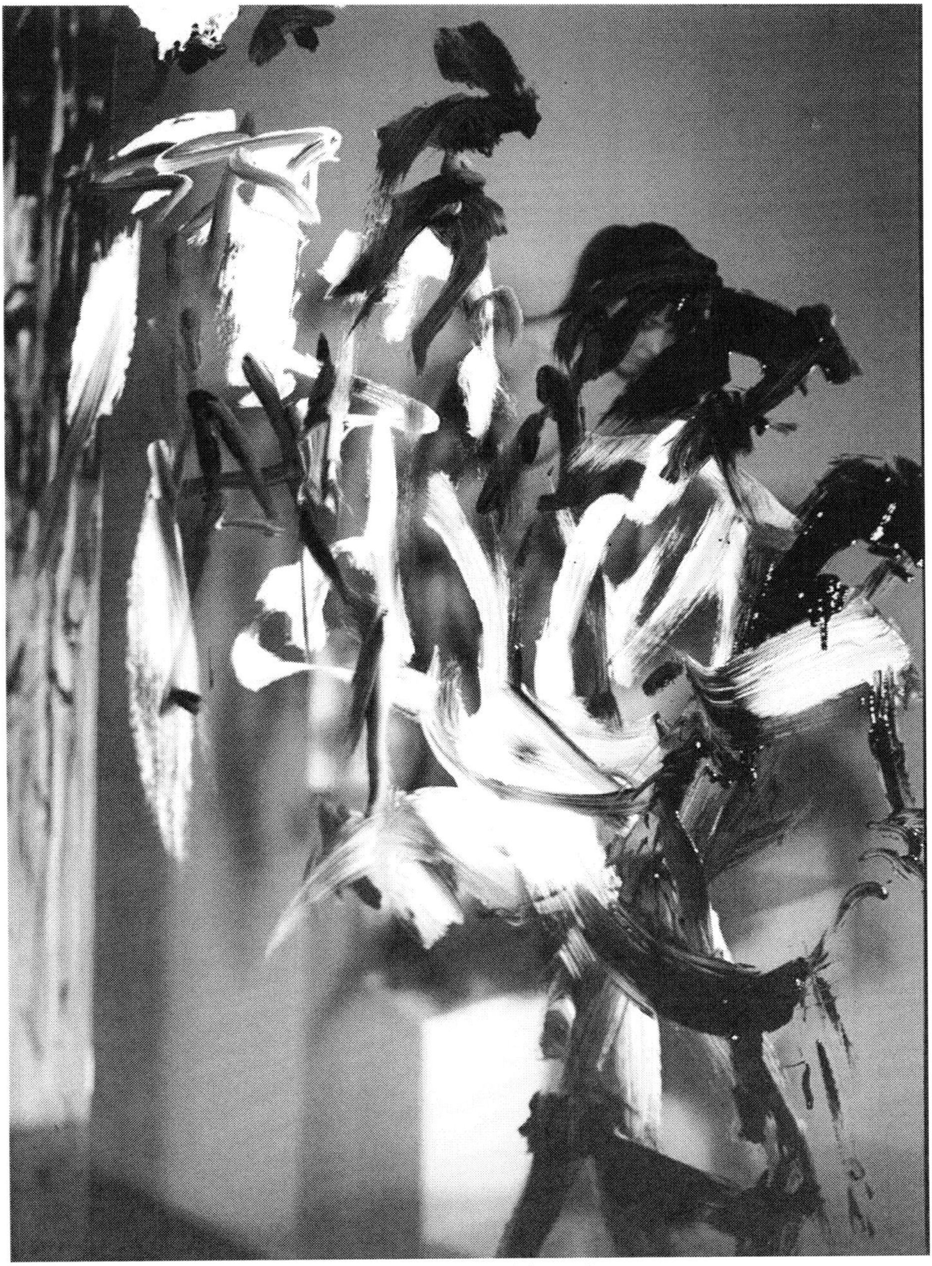

Tanya Harnett, 2008, reprinted with permission of the artist.

Figure 26. Nude Descending a Staircase.

4.2. 3D Motion Capture Technology on Visual Art

If photographic techniques tested the limits of visual culture in the early 20th century, film, video and more recent developments in computers (3-D motion capture, virtual reality, etc.) do so in ours. Cognition in these dynamic environments also relies upon the empathy of the viewers. These newer technologies are relatively easy to use and they diminish demands formerly made on content creators; simply, because video media unfolds sequentially, it is easy for viewers to believe in its verisimilitude. With respect to motion capture technology as an extension of video media, four fundamental and consequential questions arise: 1) Can 3-D technologies facilitate analysis and categorization of dynamic, time-based motions? 2) How much representative information is needed in order to evoke an event's unfolding over time (affects of human movement)? 3) What implications does the technology have for artistic image making? And, 4) can we begin to speculate about cultural context as a factor influencing gesture?

4.3. 3D Motion Capture & the Visual Arts: Theoretical Considerations

Gesture communicates information, the meaning of which is driven by empirical experience. Perceptions of gesture as an unfolding of time-based events are intrinsically integrated with the form and function of human movement. Since viewers of a 2-D image have experienced their own bodily gestures as a time-based event, they are inclined to empathize with any such implications in an artwork. The aesthetic experience and understanding of a visual image is both spatial and temporal. Hence, visual communication is linked to the personal experience of the viewer, where culture, iconography and/or visual analogy can play a role.

Motion capture provides the means to accurately record movement in 3-dimensional computer space. The technique differs significantly from actual photography. First, the information gathered is not an actual image, rather it is a collection of trajectories generated by pre-defined, user-determined points. Second, the viewer's choice of perspective is not controlled by camera placement because it can be altered after the motion has been captured. Images synthesized from motion capture trajectories may be used to analyze characteristics that are objectively representational of time and bodiliness. These may be considered signature elements of human movement.

4.4. Exemplary Experiments

Notwithstanding type of activity, trajectory-based images generated using motion capture suggested qualities that may be an aid to establishing signature elements of human movement. The uniqueness of each activity can be understood by relational emphasis on factors such as stability (balance, body position/posture), athleticism (weight, speed, power/vigor), quality of movement (control, smoothness/abruptness), and dimension of the movement (excursion). Selected examples, described below, showed how 3-D motion capture could aid discourse about perceptual characteristics of human movement. Examples are drawn from sports and dance.

Figures 27 & 28 shows trajectories associated with a side kicking technique in Tae Kwon Do. Its main characteristics were:

Stability

Trunk trajectories had relatively small, near-linear excursions and they tended to be clumped in one area of the image. The largest trajectories (legs and feet) visibly occurred on a plane rather than in a volume. Clear presence of a centre, circled by rotational trunk trajectories suggested weight, balance and a centripetal centre.

Athleticism

Long trajectory lines of the legs, particularly the feet contrasted the "low-energy", shorter lines of the trunk, lending the impression of a speed differential and a directed power. Orientation of the foot and leg trajectories was in contrast to those of the rest of the image, showing dynamism.

Quality of Movement

Continuities of trajectories were smooth, showing controlled gestures, except for acute-angle structures that occurred at the apex of the foot excursion. Abrupt changes of the direction indicated thresholds of the activity or bodily limits where energy of the movement was released. Generally compact excursions of trajectory lines showed the movement to be occurring in a limited space, suggesting bodily control.

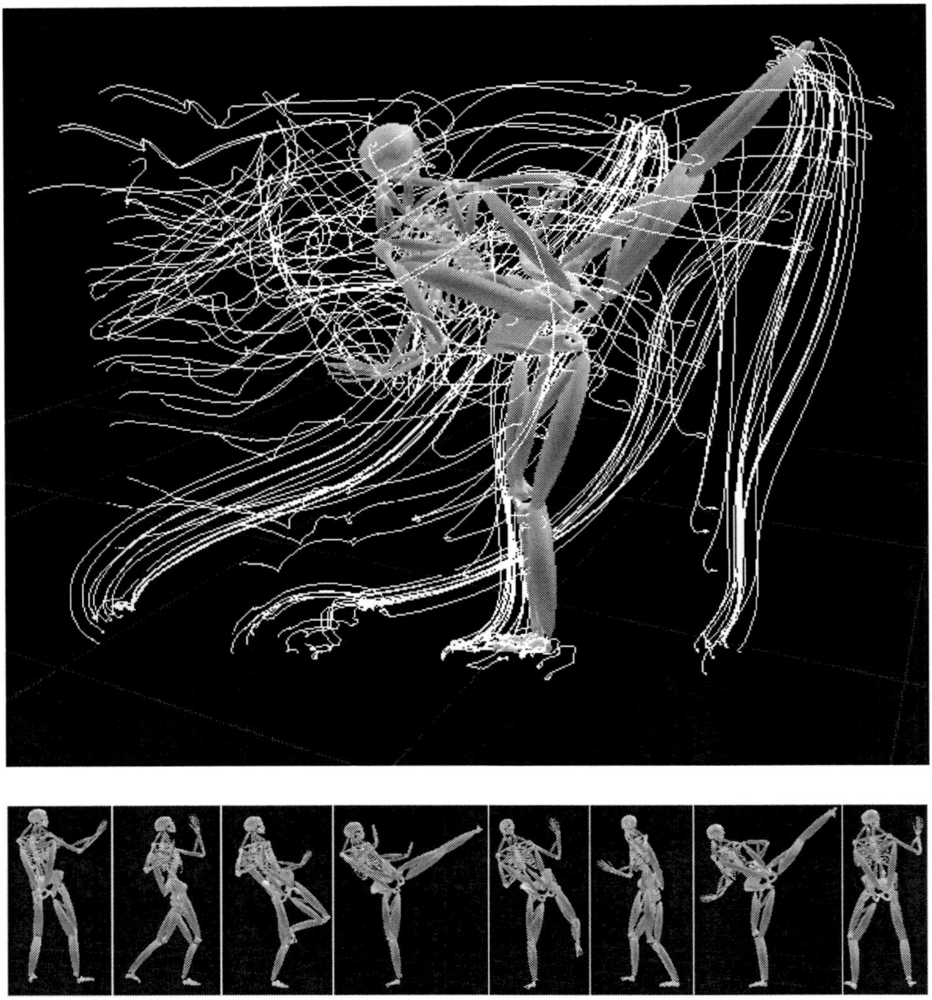

Figure 27. Side view of Tae Kwon Do side kick with body trajectory lines included (above). Sequential stop-frame reconstruction of the movement (below).

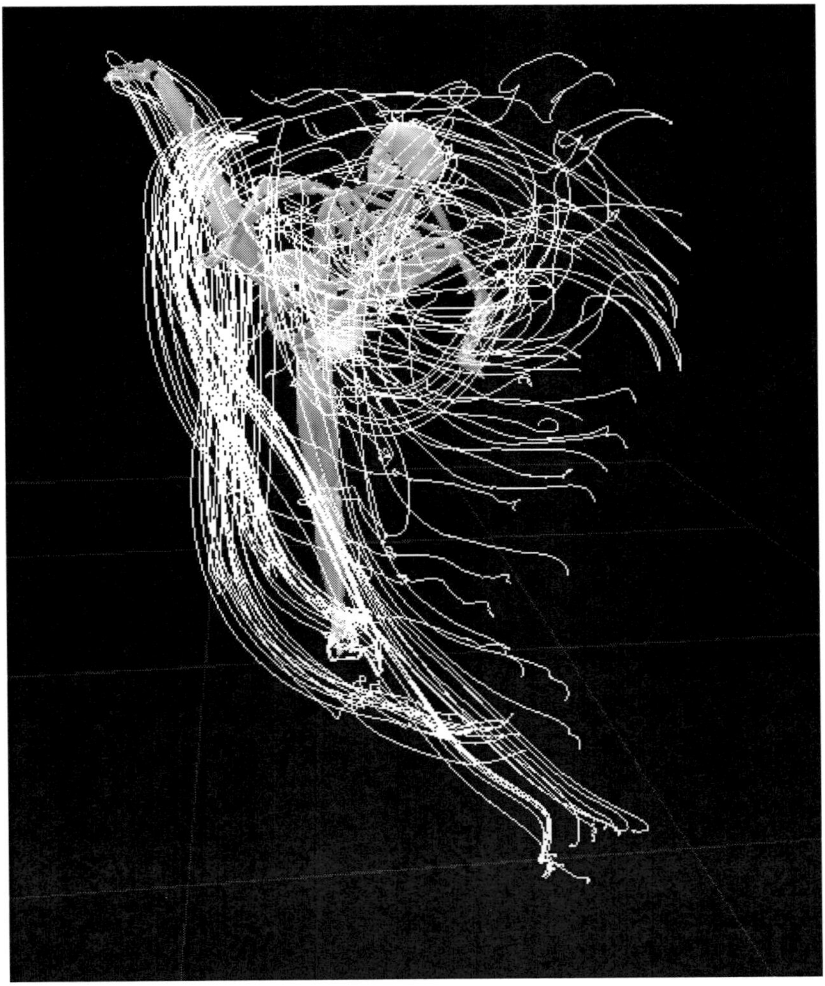

Figure 28. Tae Kwon Do side kick (front-back view) with body trajectory lines.

Figure 29 shows a soccer kick, viewed from above. Its key characteristics were:

Stability

Most trajectories appeared to be aligned in one general direction and have nearly similar lengths. Relatively parallel to one another, trajectories (particularly trunk and head) showed a state of controlled balance.

Athleticism and Dimension

Swirling trajectories of the arms were related to energy and athleticism. The double structure of these swirls indicated building and release phases of the movement. It also gave scale to the overall gesture.

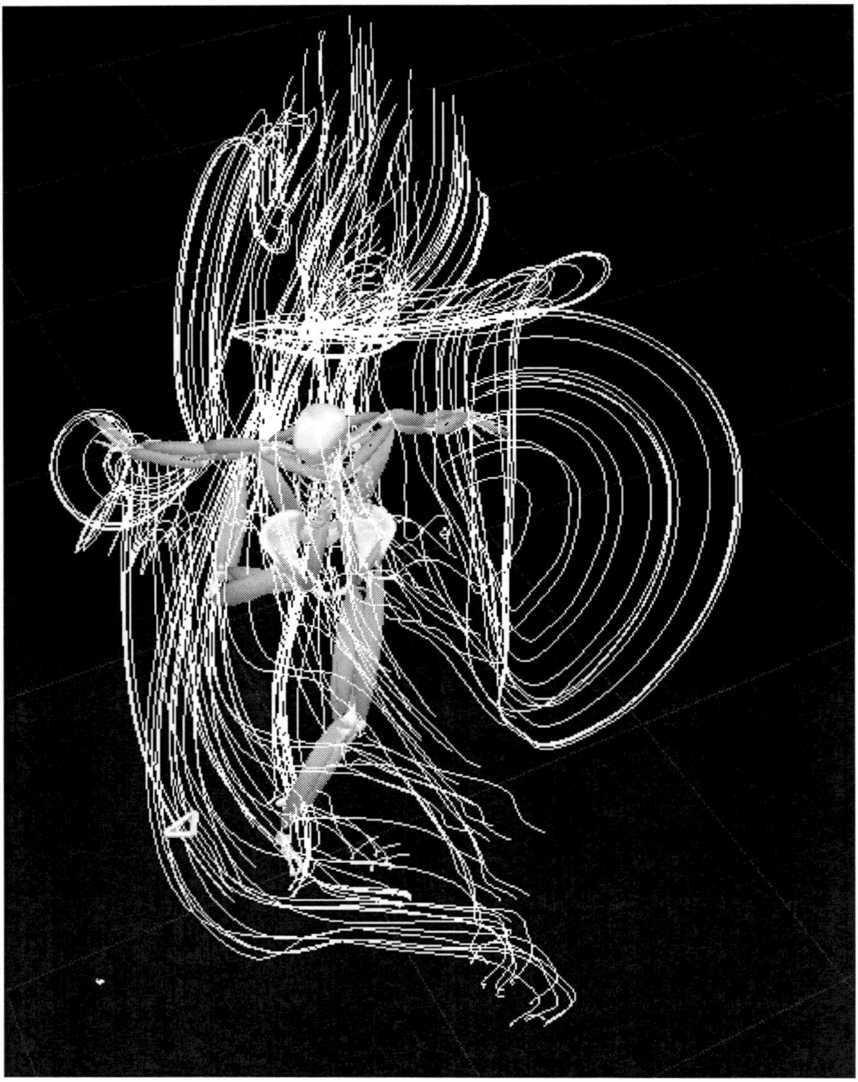

Figure 29. Soccer maximal instep kick (top view) with body trajectory lines.

Quality of Movement

Abrupt changes in trajectory direction (points of curvilinear discontinuity) showed the release of pent-up power, a quality related to dynamism.

Figure 30 shows a signature generated by a Grand Jeté from classical ballet.

Stability and Quality

A dense mass of trajectories, and a planar orientation in the central trajectories of the image (trunk) were evidence of balance and control. These contrasted trajectories (e.g. arm) where excursions had a rotational focal point (i.e. shoulder) and the regularity of the curvilinear path clearly indicated fluidity and speed. These signature characteristics can be related to balancing.

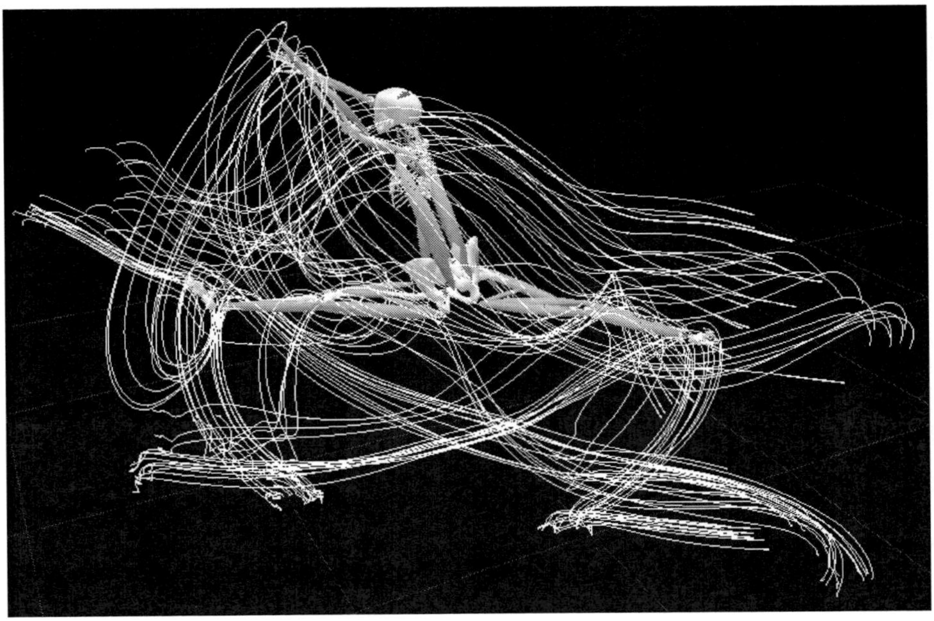

Figure 30. Ballet Grand Jété with body trajectory lines.

Figure 31 shows a fundamental skill of classical Indian dance. In Indian Dance, hand gestures and steps are executed according to highly codified traditions. Different combinations of gestures and movements provide a huge vocabulary through which stories can be depicted in surprising detail, perhaps intended to surpass the spoken word (Barboza, 2006).

Stability and Control

The clearly formal structure of trajectory lines showed controlled movement. Layered densities of line showed elements of architectural symmetry, balance, and regularity & consistency of movement. Centre of gravity was virtually fixed in a horizontal plane. Such characteristics, particularly the symmetrical, architectural references which were the result of highly codified training, could perhaps be seen to be cultural indicators.

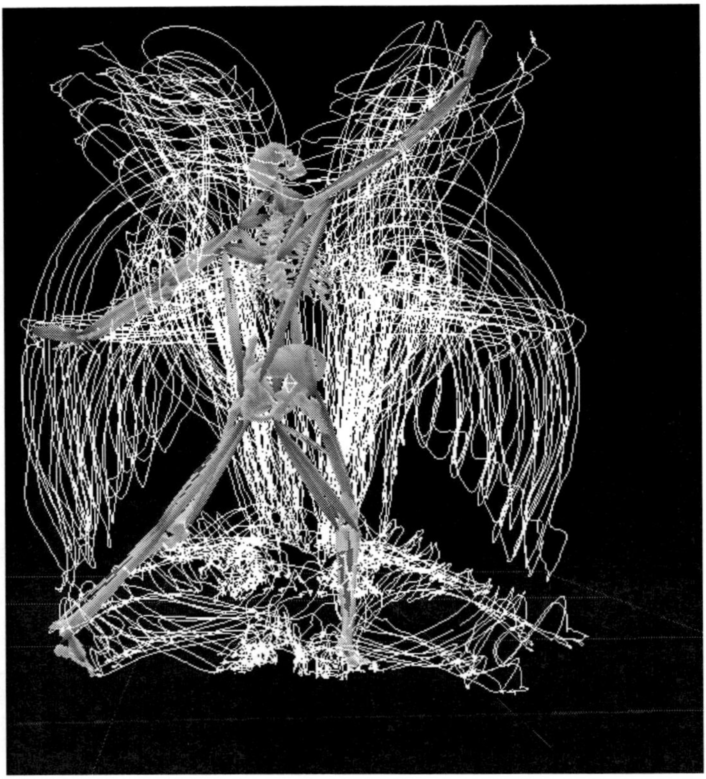

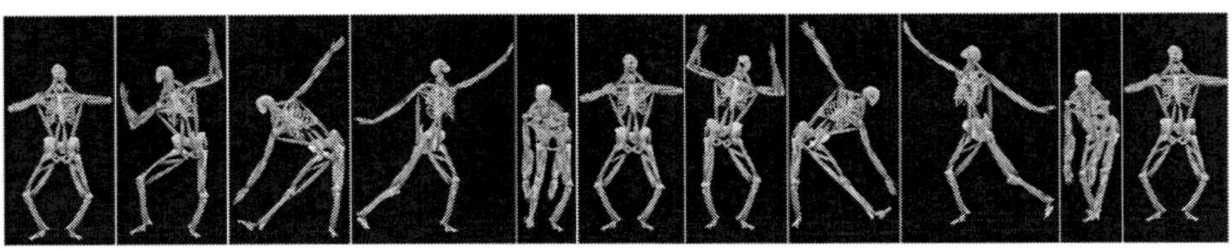

Figure 31. Indian classical dance (front View) with body trajectory lines. Sequential stop-frame reconstruction of the movement (below).

Athleticism and Quality

Continuous gesture trajectories along with no obvious points of curvilinear irregularity indicated dynamism to be less important. Energy expenditure was directed toward control rather than athleticism, which would have more indications of gathering and release. Generally compact excursions of trajectory lines suggested movements in a limited space.

Summing up, for human movement, the experiential nature of bodiliness is a link connecting the spatial and temporal. Activities such as sports, dance and performance are loaded with cultural mythos, whether it is the embodiment of the warrior class, religious symbolism or the inertia of cultural fashion. Movement signatures necessarily play a role in storytelling. This assertion acknowledges important differences between the concepts of "history" and "aetiology". Whereas history is often viewed as developmental, which should not be conflated with eschatological, aetiology investigates the conditions needed for historical change to take place. Change occurs within the context of culture – whether biologically or sociologically motivated.

Examining sports, its heritage harkens to surrogacy for warrior-class activities, emphasizing power and speed, with additional elements such as control and balance. These can be understood as signature elements of activities derived from a tradition of battle. Culturally, sports teams are an extension of the concept of championing, where selected representatives of the group pit their skills against opponents in proxy for the group as a whole. The signatures of Figures 27 to 29 reveal these very characteristics.

Performing arts activities such as dance represent traditions that can also be understood as a proxy; one for the *values* of the group heritage. In the case of activity 4, Grand Jeté, cultural references are rooted in the Court of Louis XIV, where the foundations of classical ballet originate and dance was used as a political tool. As a matter of note, when he took power after reaching the age of majority, Louis the XIV, a virtuoso dancer, performed in an allegorical ballet, dressed as the Sun God Apollo, symbolizing the rising of the sun to drive away of the darkness of the previous political regime. From this point forward, Louis XIV's noble class involved much of their efforts in learning ballet in order to imitate the noble bearing of the King (Hilton, 1997). Hence, fundamental signature elements - upright bearing, control, naturalness and demonstrated strength through elegant fluidity - become highly codified; controlled symmetrical dance structures and steps are analogous with the nobility of the King and his natural, god-given place in the universe. These elements are still present in the fundamentals of ballet today (Figure 30).

Clearly, many activities can be viewed as culturally loaded. The East Indian dance signature (Figure 31) is visually distinctive, with characteristics that display highly codified dance traditions – a stylization that captures something that is meaningful to or representative of the culture. In East Indian dance, iconography is very direct. Much is related to storytelling (religious or historical) through gesture. Formal, architectural structures, stability and control can be related to the rhetorical cohesion needed to tell these stories. Sociological and/or anthropological understanding enhanced by movement analysis technology might be one bridge in the exchange of cross-cultural ideas. Understanding our experience as a temporal bodily phenomenon creates cognitive links between information, still imagery, and cultural mythos as other kinds of metaphors for the human experience.

4.5. Other Contemporary Examples

The potential of movement analysis technology in the creation of art has been explored in several settings. Notably, Paradiso (1999) explored unconventional interfaces for individuals to interact with musical environments. In his "Brain Opera" project responsive environments were built out of "smart objects and sensitive spaces, where any kind of physical activity or motion can result in a sophisticated multimedia reaction" (Paradiso, 1999). The kinds of interfaces chosen permitted interactive musical mappings based on the intuitive responses of the participants (who were also the audience). Although the authors claim that several of the Brain Opera's goals were achieved, the experiment has essentially proven to be a "one-off". It may have been an ambitious way of "hinting at the possibilities now opening for large-scale participatory music installations," however the aim seems to have been mostly an exploration of technological capacity and the project fell short of enhancing the intensity and relevance of an artistic concept. In the study, Paradiso explained this to arise from the way in which the "project refuse(d) to fit into any clearly defined musical category". In a separate study, Paradiso et al. (2003) further examined the concept of interactive musical environments using electromagnetic tagging technology, where an ensemble of passively tagged objects was tracked in real time. These were tied to simple notes and musical effects which enabled dynamic control of musical sequences and sonic textures (Paradiso, Pardue, Hsiao, and Benbasat, 2003). Although technologies can advance installation-based interactive music environments, their communicative efficacy still rests in the ears of the audience.

In a multimedia project incorporating music and dance. Bevilaqua et al. (2001) used a 3D motion capture system to produce animations from dance and, more unusually, generate a soundtrack from the dancer's movements. Motion capture data was used to both trigger and modify the timbre of sounds. The project was exploratory, challenging traditional boundaries between the arts (music, dance, studio art and drama) and sciences (motion analysis technology). The authors define their thesis as follows:

> "In electronic music, the relationship between sound generation and touch/gesture is artificially defined, contrary to the way in which sound is produced in acoustic instruments. In fact, the possibility of designing the interface between the gesture (or the touch) and the sound is a fascinating feature of digital music. Most electronic music controllers that have been created are based on existing acoustic instruments (for example the piano keyboard). Such electronic controllers have the obvious advantage to be used relatively easily by 'traditionally' trained musicians" (Bevilacqua, Naugle, & Valverde, 2001).

In summary, such research consistently points to the need and desire for greater collaboration between artists and scientists. In the given examples, technology provides analytic tools and a means to characterize intersections between the body and artistic creation. Biomechanical signals link the artistic conceptualization and its interface with our perceptions. Such explorations are just starting. And, while the benefit may currently seem a bit one-sided – artists taking advantage of technology for the betterment of their art – the benefit really flows both ways. Technology is in a consistent state of rapid change. Artists' demands can be a catalyst for improving technology. For example, in the case of 3-D motion capture, some of the first demands for both video and sound to be synchronized with the 3-D data came from Arts researchers (VICON motion capture). This development is ubiquitous in the industry today.

4.6. In Conclusion and Coming Full Circle

In terms of the technology of biomechanics research, there continues to be significant unexplored potential. Throughout history, each set of tools has ushered its own set of changes into the way people interact with their environment and generate understanding. Technology may allow complex mapping functions to be interpreted very sophisticated ways. But, it is sometimes more interesting and useful to consider the artistic gesture, the potential for expression, or the aesthetic design inherent in the digital data generated by it. This kind of thought is a central tenant of the artistic process. And, to reconsider the first question posed in Chapter 1.1 of this book, it is one of the best reasons to “bother with Arts Biomechanics.”

ACKNOWLEDGEMENTS

The creation of this book was financially supported by NSERC (National Sciences and Engineering Research Council of Canada), Westgrid/Canada, CREDO (University of Letbridge). Further, the authors wish to acknowledge and thank the staff in the Office of Research Service/University of Lethbridge for their dedicated assistance.

References

Administration, F. A. (1996). Dynamic Evaluation of Seat Restraint Systems &Occupant Protection on Transport Airplanes. Retrieved. from.

Bach, C. P. E. (1759). *Essay on the True Art of Playing Keyboard Instruments* (W. J. Mitchell, Trans.). New York: W. W. Norton & Co.

Baillot, P. (1835). *L'art du violon* (L. Goldberg, Trans.). Paris: Northwestern University Press.

Ballreich, R., & Baumann, W. (1996). Grundlagen der Biomechanik des Sports (The basics of biomechanics in sports). Stuttgart: Enke Verlag.

Ballreich, R., & Kuhlow, A. (1980). *Beiträge zur Biomechanik des Sports*. Schorndorf: Karl Hofmann.

Barboza, K. (2006). Repräsentative und symbolische Zeichen des klassischen Tanzes in Indien und in Europa: Eine interkulturelle, vergleichende Analyse des indischen bharatanāyam und des französischen Balletts. University of Münster, Münster.

Barth, T. (2000). An AIR passenger airbag system for transport aircraft. Retrieved. from.

Bevilacqua, F., Naugle, L., & Valverde, I. (2001). *Virtual dance and music environment using motion capture.* Paper presented at the IEEE Multimedia Technology and Applications Conference, Irvine.

Biral, D., Jakubiec-Puka, A., Ciechomska, I., Sandri, M., Rossini, K., Carraro, U., et al. (2000). Loss of dystrophin and some dystrophin-associated proteins with concomitant signs of apoptosis in rat leg muscle overworked in extension. *Acta Neuropathologica, 100*(6), 618-626.

Brandser, E. A., Elkhoury, G. Y., Kathol, M. H., Callaghan, J. J., & Tearse, D. S. (1995). Hamstring Injuries - Radiographic, Conventional Tomographic, Ct, And Mr-Imaging Characteristics. *Radiology, 197*(1), 257-262.

Braun, M. (1992). Picturing Time: The Work of Etienne-Jules Marey (1830-1904). Chicago: University of Chicago Press.

Brown, A. N. (1997). Musculoskeletal misuse among youth symphony string players. *Medical Problems of Performing Artists, 12*, 15-18.

Cavanagh, P. (1990). The mechanics of distance running: a historical perspective. In P. Cavanagh (Ed.), *Biomechanics of Distance Running* (pp. 135-164). Champaign: Human Kinetics.

Chesky, K., Kondraske, G., Henoch, M., Hipple, J., & Rubin, B. (2002). Musicians' health. In R. Colwell & C. Richardson (Eds.), *The new handbook of research on music teaching and learning*. New York: Oxford University Press.

Chockley, C. (2008). Ground Reaction Force Comparison Between Jumps Landing on the Full Foot and Jumps Landing en Pointe in Ballet Dancers. *Journal of Dance Medicine and Science, 12*(1).

Dawson, D. M., Hallett, M., & Millender, L. H. (1990). *Entrapment Neuropathies*. Boston: Little, Brown & Co.

Dawson, W. J. (2003). The bibliography of performing arts medicine: A five-year retrospective review. *Medical Problems of Performing Artists, 18*(1), 27-32.

Dawson, W. J. (2007). The bibliography of performing arts medicine - A 10 year retrospective review (Part II). *Medical Problems of Performing Artists, 22*(4), 153-159.

Devroop, K., & Chesky, K. (2002). Comparison of biomechanical forces generated during trumpet performance in contrasting settings. *Medical Problems of Performing Artists, 17*(4), 149-154.

Einstein, A. (1931). The World as I See It. Retrieved January 9, 2009, from http://www.aip.org/history/einstein/essay.htm

Evans, B. (2003). How I Survived My Dance Training. *Medical Problems of Performing Artists, 18*(4), 137-140.

Fotheringham W. (2002). *Put Me Back on My Bike: In Search of Tom Simpson.* New York: Yellow Jersey Press.

Fry, H. J. H. (1986). Overuse syndrome of the upper limb in musicians. *Medical Journal Austrilia, 144*, 182-185.

Fry, H. J. H. (1987). Prevalence of overuse (injury) syndrome in Australian music schools. *British Journal of Internal Medicine, 44*, 35-40.

Fry, H. J. H. (1988). Patterns of over-use seen in 658 affected instrumental musicians. *International Journal of Music Education, 11*, 3-16.

Fry, H. J. H., Ross, P., & Rutherford, M. (1988). Music-related overuse in secondary schools. *Medical Problems of Performing Artists, 3*, 133-134.

Gadd, C. (1966). *Use of a weighted impulse criterion for estimating injury hazard.* Paper presented at the 10th Stapp Car Crash Conference, New York.

Gadd, C. M. (1966). *Use of a weighted impulse criterion for estimating injury hazard.* Paper presented at the 10th Stapp Car Crash Conference, New York.

Hagglund, K. L. (1996). A comparison of the physical and mental practices of music students from the New England Conservatory and Boston University Music School. *Medical Problems of Performing Artists, 11*, 99-107.

Hartsell, H. D., & Tata, G. E. (1991). A retrospective of music-related musculoskeletal problems occurring in undergraduate music students. *Pysiotherapy Canada 43*(1), 13-18.

Hennig, E. M., & Lafortune, M. A. (1991). Relationships between ground reaction force and tibial bone acceleration parameters. *International Journal of Sport Biomechanics, 7*, 303-309.

Hilton, W. (1997). Dance and Music of Court and Theatre: selected Writings of Wendy Hilton. Stuyvesant NY: Pendragon Press.

Hodgson, P. (1958). *Motion Study and Violin Bowing*. Urbana: American String Teachers Association.

Hunter, S., Jacobus, J., & Wheeler, D. (2004). *Modern Art, Third Edition Revised and Expanded*. New York: Prentice-Hall.

Kanbe, K., Hasegawa, A., Takagishi, K., Shirakura, K., Nagase, M., Yanagawa, T., et al. (1998). Analysis of muscle bioenergetic metabolism in rabbit leg lengthening. *Clinical Orthopaedics and Related Research*(351), 214-221.

Kaplinski, V. (2003). *Prevention of injuries among keyboard players*. Paper presented at the Twenty-First Annual Symposium on Medical Problems of Musicians and Dancers.

Kline, F. (1950). Chief. Retrieved March 22, 2008, 2008, from http://www.moma.org/collection/browse_results.php?object_id=78319

Kolneder, W. (1993). The Amadeus Book of the Violin - Construction, History, and Music. Oregon: Amadeus Press.

Komulainen, J., Takala, T. E. S., Kuipers, H., & Hesselink, M. K. C. (1998). The disruption of myofibre structures in rat skeletal muscle after forced lengthening contractions. *Pflugers Archiv-European Journal of Physiology, 436*(5), 735-741.

Kosiak, M. (1959). Etiology and Pathology of ischemic ulcers. *Archives of Physical Medicine, 40*, 62-69.

Koskinen, S. O. A., Ahtikoski, A. M., Komulainen, J., Hesselink, M. K. C., Drost, M. R., & Takala, T. E. S. (2002). Short-term effects of forced eccentric contractions on collagen synthesis and degradation in rat skeletal muscle. *Pflugers Archiv-European Journal of Physiology, 444*(1-2), 59-72.

Kuipers, H. (1998). Training and overtraining: An introduction. *Medicine & Science in Sports & Exercise, 30*, 1137-1139.

Lehmann, A. C., & Davidson, J. W. (2002). Taking an acquired skills perspective. In R. Colwell & C. Richardson (Eds.), *The new handbook of research on music teaching and learning*. New York: Oxford University Press.

Levitin, D. (2006). *This is Your Brain on Music*. New York: Plume.

Lockwood, A. H. (1988). Medical problems in secondary school-aged musicians. *Medical Problems of Performing Artists, 3*, 129-132.

Lundberg, A. (1996). On the use of bone and skin markers in kinematics research. *Human Movement Science, 15*(3), 411-422.

Magill, R. A. (2001). *Motor learning concepts and applications* (6 ed.). Boston: McGraw-Hill.

May B., & Shippen J. (2008). Biomechanical modelling system to solve muscular loading distribution. Paper presented at the Southeastern Meeting of the American Society of Biomechanics, Birmingham, AL.

McLean, A., & Anderson, R. (1997). Biomechanics of closed head injuries. In P. Reilly & R. Bullock (Eds.), *Head Injury* (pp. 25-37). London: Chapman & Hall.

Medved, V. (2001). *Measurement of Human Locomotion*. Boca Raton: CRC Press.

Middlestadt, S. E., & Fischbein, M. (1989). The prevelance of severe musculoskeletal problems among male and female symphony orchestra string players. *Medical Problems of Performing Artists, 4*, 41-48.

Moyse, M. (1973). The Flute and its Problems; Tone Development Through Interpretation for the Flute. Tokyo: Muramatsu Gakki Hanbai Co.,Ltd.

Mozart, L. (1756). *A treatise on the Fundamental Principles of Violin Playing* (E. Knocker, Trans.). Oxford: Oxford University Press.

Nemoto, K., & Arino, H. (2007). Hand and upper extremity problems in wind instrument players in military bands. *Medical Problems of Performing Artists, 22*(2), 67-69.

Nicol, K. (1996). Velocity vs. duration of competition - A tool for characterizing sport events, internal load, and limiting factors. Paper presented at the 9th Biennial Conference, Canadian Society for Biomechanics, Vancouver.

Nicol, K. (1998). Gesetzmäßigkeiten von Grenzleistungen. *Leistungssport, 5* 28-35.

Palac, J. A. (1992). Violin bowing technique: an analysis of contemporary pedagogical literature according to principles of human movement. *Medical Problems of Performing Artists, 7*(1), 30-34.

Paradiso, J. A. (1999). The brain opera technology: New instruments and gestural sensors for musical interaction and performance. *Journal of New Music Research, 28*(2), 130-149.

Paradiso, J. A., Pardue, L. S., Hsiao, K. Y., & Benbasat, A. Y. (2003). Electromagnetic tagging for electronic music interfaces. *Journal of New Music Research, 32*(4), 395-409.

Parlitz, D., Peschel, T., & Altenmuller, E. (1998). Assessment of dynamic finger forces in pianists: Effects of training and expertise. *Journal of Biomechanics, 31*(11), 1063-1067.

Peterson, L., & Renström, P. (1987). *Verletzungen im sport (Injuries in Sports).* Köln: Deutcher Ärte-Verlag.

Plagenhoef, S. (1971). *Patterns of Human Movement: a cinematographic analysis*. Englewood Cliffs: Prentice-Hall.

Quantz, J. J. (1752). *Essay of a Method for Playing the Transverse Flute* (R. E. R., Trans.). London: faber and faber.

Quinn, W. (1990). The Paradox of the Self-Torturer. *Philosophical Studies,* 59, 79-90.

Rabuffetti, M., Converti, R. M., Boccardi, S., & Ferrarin, M. (2007). Tuning of the violin-performer interface: An experimental study about the effects of shoulder rest variations on playing kinematics. *Medical Problems of Performing Artists, 22*(2), 58-66.

Rubin, S. J., Feldman, F., Staron, R. B., Zwass, A., Totterman, S., & Meyers, S. P. (1995). MAGNETIC-RESONANCE-IMAGING OF MUSCLE INJURY. *Clinical Imaging, 19*(4), 263-269.

Schwartz, B. (1983). *Great masters of the violin*. New York: Simon and Schuster.

Shan, G. (2005). Comparison of repetitive movements between ballet dancers and martial artists ¾ risk assessment of muscle overuse injuries and prevention strategies. *Research in Sports Medicine, 13*(1), 63-76.

Shan, G., Visentin, P., & Harnett, T. (2010). A Novel Use of 3D Motion Capture: Creating Conceptual Links between Technology and Representation of Human Gesture in the Visual Arts. *Leonardo, 43*(1).

Shan, G. B. (1999). Ein biomechanisches Modell auf Basis von individuellen anthropometrischen Daten fuer das Bewegungslernen der Flugphasen bei sportlichen Bewegungsablaeufen (A biomechanical model for motor learning of air-borne phases based on individual anthropometry). Müster: Lit Verlag.

Shan, G. B., & Bohn, C. (2003). Anthropometrical data and coefficients of regression related to gender and race. *Applied Ergonomics, 34*(4), 327-337.

Shan, G.B., & Visentin, P. (2010). A Prisoner's dilemma with Asymmetrical payoffs: Revealing the challenges faced by performing arts health and wellness practitioners. Medical Problems of Performing Artists.

Shan, G. B., & Visentin, P. (2003). A quantitative three-dimensional analysis of arm kinematics in violin performance. *Medical Problems of Performing Artists, 18*(1), 3-10.

Shan, G. B., Visentin, P., & Schultz, A. (2004). Multidimensional signal analysis as a means of better understanding factors associated with repetitive use in violin performance. *Medical Problems of Performing Artists, 19*(3), 129-139.

Shan, G. B., Visentin, P., Wooldridge, L., Wang, C. D., & Connolly, D. (2007). A frequency-based characterization of spiccato bowing in violin performance. *Perceptual and Motor Skills, 105*(3), 1027-1051.

Shippen, J., & May, B. (2008). *Whole body biomechanical analysis and muscle loading of Irish dancers.* Paper presented at the International Association for Dance Medicine and Science, 18th Annual Meeting, Cleveland, OH.

Shippen, J., & May, B. (2009). An analysis procedure for the calculation of muscle loading and joint contact forces in dancers. *Journal of Dance Medicine and Science.*

Smith, H. K., Plyley, M. J., Rodgers, C. D., & McKee, N. H. (1997). Skeletal muscle damage in the rat hindlimb following single or repeated daily bouts of downhill exercise. *International Journal of Sports Medicine, 18*(2), 94-100.

Solnit, R. (2003). River of Shadows: Eadward Muybridge and the Technological Wild West. New York: Viking Penguin.

Solomon, R., & Solomon, J. (2004). Publications in dance medicine and science: A bibliographer's perspective. *Medical Problems of Performing Artists, 19*(4), 167-169.

Stauber, W. T., & Willems, M. E. T. (2002). Prevention of histopathologic changes from 30 repeated stretches of active rat skeletal muscles by long inter-stretch rest times. *European Journal of Applied Physiology, 88*(1-2), 94-99.

Steinbach, L. S., Fleckenstein, J. L., & Mink, J. H. (1994). Magnetic-Resonance-Imaging Of Muscle Injuries. *Orthopedics, 17*(11), 991-999.

Sundar, M., & Carty, H. (1994). Avulsion Fractures Of The Pelvis In Children - A Report Of 32 Fractures And Their Outcome. *Skeletal Radiology, 23*(2), 85-90.

Taylor, J. (1996). Intensity and athletic performance. In J. Van Raalte & B. Brewer (Eds.), *A practitioner's guide to sport and exercise psychology.* Washington: American Psychological Association.

Thompson, J. L., Balog, E. M., Fitts, R. H., & Riley, D. A. (1999). Five myofibrillar lesion types in eccentrically challenged, unloaded rat adductor longus muscle - A test model. *Anatomical Record, 254*(1), 39-52.

Tromlitz, J. G. (1791). *The Virtuoso Flute Player* (A. Powell, Trans.): Cambridge.

Tulchinsky, E. (1995). A Biomechanical Motion Analysis Of The Violinists Bow Arm (VOL 9, PG 119, YR 1994). *Medical Problems of Performing Artists, 10*(1), 10-10.

Tulchinsky, E., & Riolo, L. (1994a). A biomechanical motion analysis of the violinist's bow arm. *Medical Problems of Performing Artists, 9*(4), 119-124.

Tulchinsky, E., & Riolo, L. (1994b). A Biomechanical Motion Analysis Of The Violinists Bow Art. *Medical Problems of Performing Artists, 9*(4), 119-124.

Uchiyama, Y., Tamaki, T., & Fukuda, H. (2001). Relationship between functional deficit and severity of experimental fast-strain injury of rat skeletal muscle. *European Journal of Applied Physiology, 85*(1-2), 1-9.

USA. Airworthiness Standards: Transport Category Airplanes. Retrieved. from.

USGovernment. Airworthiness Standards: Transport Category Airplanes. Retrieved. from.

Vijayan, K., Thompson, J. L., Norenberg, K. M., Fitts, R. H., & Riley, D. A. (2001). Fiber-type susceptibility to eccentric contraction-induced damage of hindlimb-unloaded rat AL muscles. *Journal of Applied Physiology, 90*(3), 770-776.

Virgilio, F. F., Chiara, M., Emilia, B., Paolo, P., & Sforza, C. (2007). Three-Dimensional Analysis of Hand and Finger Movements during Piano Playing. *Medical Problems of Performing Artists, 22*(1), 18-23.

Visentin, P., Shan, G., & Wasiak, E. B. (2008). Informing music teaching and learning using movement analysis technology. *International Journal of Music Education, 26*(1), 73-87.

Visentin, P., & Shan, G. B. (2003). The kinetic characteristics of the bow arm during violin performance: An examination of internal loads as a function of tempo. *Medical Problems of Performing Artists, 18*(3), 91-97.

Visentin, P., & Shan, G. B. (2004). An innovative approach to understand overuse injuries: Biomechanical modeling as a platform to integrate information obtained from various analytic tools. *Medical Problems of Performing Artists, 19*(2), 90-96.

Wilson, F. (1986). *Tone deaf and all thumbs?* New York: Viking Press.

Winter, D. A. (1990). *Biomechanics and Motor Control of Human Movement.* New York: Wiley.

Wolff, D. (2008). *Zur Optimierbarkeit von Klaviertechnik: Erfassung und Evaluation von Bewegungsabläufen und Übestrategien.* Mainz: Schott Music GmbH & Co KG.

Zaza, C. (1992). Playing-related health problems at a Canadian music school. *Medical Problems of Performing Artists, 7*, 48-51.

Zaza, C. (1998). Playing-related musculoskeletal disorders in musicians. *Canadian Medical Association Journal, 158*, 1019-1025.

INDEX

A

Acceleration, 27, 34-39, 46-49
Aesthetic(s), 7, 11, 14, 18, 25, 40-41, 70, 79
Anthropometry, 20, 23, 25, 29, 36, 46, 56
Arts Biomechanics, 7-10, 17-43, 55-57, 78
Associative, 20, 24-25
Athletiscism, 65, 71-77
Autonomous, 24-25

B

Ballet, 11, 28, 33, 75-77
Ballreich, 9, 21, 23-25, 45
Biological Reaction, 46-50
Biomechanical modeling, 27-29, 35-37, 44-46, 62-64
Body height, 39
BH (see Body height)

C

Cognitive, 20, 24-26, 34, 78
Characterization of motor skill, 23

D

Dance (also see Ballet, Indian Dance), 14, 21, 33, 39, 69, 75-79
Damage, 47-49, 52-53, 69

E

Efficiency, 19, 33, 38-39, 41, 59
Effectiveness, 33, 36, 39, 41, 59
Elbow height, 39
Electromyography, 27, 30-33
EMG (see Electromygraphy)
EMG envelope, 30-31
Exposure (photographic), 26, 65-67

F

Force, 20, 27, 29-30, 33-35, 46-49, 59, 65
Filter, 30
Force measurement, 27, 29, 33-35
Fry, 8, 19, 43-44

G

Grand Jété, 75
Ground reaction, 33-34

H

Harnett, 68-69

I

Injury risk, 10, 43, 51-56
Injury prevention, 7, 25
Indian Dance, 75-78

J

Joint angle, 37

K

Kick (also see Soccer, Tae Kwon Do), 71-74
Kinematics, 27

L

Loading, 30-31, 60-61

M

Magill, 20-21, 24-25
Marker-based teaching, 23, 27
Markers, 28-29, 36-39
Maximum voluntary contraction, 31-33
Mechanical work, 47
Mechanics, 14
Micro-damage (see Micro-trauma)
Micro-trauma, 44-45, 52-54, 61-63
Motion analysis, 23, 35, 55, 64, 68, 79
Motor control, 9, 17-25, 29-30, 35-38, 44, 55-57
Motor learning, 7, 9, 18-25, 39
Movement analysis (see Motion analysis)
Movement quality, 71, 75, 77
MRI, 61-62
Multidisciplinary, 10, 21, 27, 56
Muscle activity, 29-31, 35, 46-47, 54, 61
Music performance, 55-57
Music teaching (see Pedagogy)
Muybridge, 65-66
MVC (please see Maximum voluntary contration)

N

Neural control (see Motor control)
Normalization, 31

O

Optimization, 7, 38, 55-56
OS (see Overuse syndrome)
Overuse syndrome, 43-44, 55-57

P

Parameter of biological reaction, 49-54
PBR (see Parameter of biologica reaction)
PD (see Prisoner's dilemma)
Pedagogy, 9, 15, 17-23, 36-40, 56
Posture, 14-15
Photograpy, 65-68
Physiological limits, 15, 21, 31, 38, 44 49, 51, 56-57, 71
Prisoner's dilemma, 57-58
Pysiological tolerance, 44-45, 51

Q

Quantitative analysis, 59
Qunn's scenario, 58

R

Range of motion, 37, 43
ROM (see Range of motion)
Repetitive Stress Injuries (also se Overuse syndrome), 24, 39 43-47, 5 54
Repetitive work, 19, 61
RSI (see Repetitive Stress Injuries)

S

Severini, 67
Signature of human movement, 70, 7 77-78
Skill Analysis (also see Motion analysis 10
Skill Acquisition, 7, 9, 14, 18-20, 26, 3 41, 56
Skill transfer, 23-24
Soccer, 74

tability, 71, 73, 75-76, 78
ynchronized measurement, 34-5

T

'ae Kwon Do, 71-73
-D Motion capture, 27-28
'ime, 8, 24, 34, 39, 44, 47-48, 50-53, 56, 67, 70
'empo, 37
-D image, 65, 70

V

Violin bowing, 14, 36-37, 39
Visual Art, 65-76
Velocity, 35, 47-48

W

Work, 46-48

Z

Zaza, 8, 19, 43